Green Energy and Technology

For other titles published in this series, go to
http://www.springer.com/series/8059

Vivian W. W. Yam
Editor

WOLEDs and Organic Photovoltaics

Recent Advances and Applications

Editor
Prof. Vivian W.W. Yam
Department of Chemistry
The University of Hong Kong
Pokfulam Road
Hong Kong
China
wwyam@hku.hk

ISSN 1865-3529 e-ISSN 1865-3537
ISBN 978-3-642-14934-4 e-ISBN 978-3-642-14935-1
DOI 10.1007/978-3-642-14935-1
Springer Heidelberg Dordrecht London New York

© Springer-Verlag Berlin Heidelberg 2010
This work is subject to copyright. All rights are reserved, whether the whole or part of the material is concerned, specifically the rights of translation, reprinting, reuse of illustrations, recitation, broadcasting, reproduction on microfilm or in any other way, and storage in data banks. Duplication of this publication or parts thereof is permitted only under the provisions of the German Copyright Law of September 9, 1965, in its current version, and permission for use must always be obtained from Springer. Violations are liable to prosecution under the German Copyright Law.

The use of general descriptive names, registered names, trademarks, etc. in this publication does not imply, even in the absence of a specific statement, that such names are exempt from the relevant protective laws and regulations and therefore free for general use.

Cover design: WMXDesign GmbH, Heidelberg, Germany

Printed on acid-free paper

Springer is part of Springer Science+Business Media (www.springer.com)

Preface

A major global issue that the world is facing today is the upcoming depletion of fossil fuels and the energy crisis. In 1998, the global annual energy consumption was 12.7 TW; of which 80% was generated from fossil fuels. This also translates into huge annual emissions of CO_2 that leads to massive environmental problems, particularly the global warming, which could be disastrous. Future global annual energy needs are also estimated to rise dramatically. A major challenge confronting the world is to find an additional 14–20 TW by 2050 when our energy reserves based on fossil fuels are vanishing. The massive demand for energy would require materials and/or processes that would help to provide new sources of clean renewable energy or to develop processes that would harvest energy or to better utilize energy in an efficient manner. The present monograph, WOLEDs and Organic Photovoltaics – Recent Advances and Applications, focuses on a very important and timely subject of topical interest that deals with the more efficient use of energy through white organic light-emitting diodes (WOLEDs) for solid-state lighting and the development of clean sources of renewable energy through the harvesting of light energy for conversion into electrical energy in organic photovoltaics. While LED solid-state lighting and photovoltaics have been dominated by inorganic semiconductor materials and silicon-based solar cells, there have been growing interests in the development of WOLEDs and organic photovoltaics. In spite of the formidable challenges that confront scientists in developing high-efficiency WOLEDs and organic solar cells, the advantages and rewards for success are substantial and may offer opportunities that are simply too great to be ignored.

The present monograph was a result of the WOLEDs and Organic Photovoltaics Workshop that was held at The University of Hong Kong (HKU) in January 2009, jointly organized by the Strategic Research Theme (SRT) on Molecular Materials and the Initiative on Clean Energy and Environment (ICEE) at HKU.

The SRT on Molecular Materials is one of the identified strategic research themes of the University, which aims to focus on the development of molecular materials with functional properties, with special emphasis on organic optoelectronics, particularly in the areas of organic and metal-organic light-emitting materials, organic light-emitting devices (OLEDs), organic thin film transistors (OTFTs), organic photovoltaics (OPVs), and dye-sensitized solar cells (DSSCs). Molecular materials with other optoelectronic, photonic, and electronic functions will also be explored. Our goal is to establish strong interdisciplinary team efforts among chemists, physicists, and engineers to work in this area to seek basic knowledge in the structure–property relationships of new patentable molecular materials and their functions, and to understand the underlying principles and science of their optoelectronic and device properties.

The ICEE was formally set-up in the Spring of 2008 with funding from the University's Development Fund to support targeted research in Clean Energy. It was borne out of a coalition of senior HKU faculty keenly aware of the increasing priority of clean energy over the coming decades and the current research strengths and proven track record of HKU researchers. These strengths range from materials science through engineering, electronics, and building-related applications to policy and economic issues. The ICEEs mission is to foster and advance interdisciplinary clean energy research at HKU and to make HKU a focal point of intellectual and academic endeavor in clean energy research in Hong Kong, China, and the world.

Through workshops like this one at HKU, it is our intention to bring together the world's leading authorities on selected topics and to publish the delivered papers in a monograph. The significance of the research topic in the development of clean energy technologies will be clearly linked.

I would like to take this opportunity to thank all the contributors to this monograph for their efforts and for their contributions to the success of the monograph and the WOLEDs and Organic Photovoltaics Workshop in Hong Kong. I would like to thank Professor Felix F. Wu of ICEE for his kind suggestion and idea of asking me to coordinate and organize the publication of a monograph related to the Workshop. I also would like to thank Springer, especially Dr. Marion Hertel and Beate Siek, for their belief in me and the project and for their strong support to make this project possible. Dr. Maggie M.Y. Chan is also gratefully acknowledged for her kind assistance in the preparation of the monograph.

Both Professor Wu and I would like to thank the SRT on Molecular Materials and ICEE of HKU and the Philip Wong Wilson Wong Endowed Professorship Fund for their generous support.

Hong Kong, China   Vivian Wing-Wah Yam
                      Philip Wong Wilson Wong Professor in Chemistry and Energy
                                            Convenor, SRT on Molecular Materials
                                  Organizer, WOLEDs and Organic Photovoltaics Workshop

Contents

Overview and Highlights of WOLEDs and Organic Solar Cells: From Research to Applications 1
Maggie Mei-Yee Chan, Chi-Hang Tao, and Vivian Wing-Wah Yam

White-Emitting Polymers and Devices 37
Hongbin Wu, Lei Ying, Wei Yang, and Yong Cao

Phosphorescent Platinum(II) Complexes for White Organic Light-Emitting Diode Applications 79
Chi-Chung Kwok, Steven Chi-Fai Kui, Siu-Wai Lai, and Chi-Ming Che

Solid-State Light-Emitting Electrochemical Cells Based on Cationic Transition Metal Complexes for White Light Generation 105
Hai-Ching Su, Ken-Tsung Wong, and Chung-Chih Wu

Horizontal Molecular Orientation in Vacuum-Deposited Organic Amorphous Films 137
Daisuke Yokoyama and Chihaya Adachi

Recent Advances in Sensitized Solar Cells 153
Arthur J. Frank

Implications of Interfacial Electronics to Performance of Organic Photovoltaic Devices 169
M.F. Lo, T.W. Ng, M.K. Fung, S.L. Lai, M.Y. Chan, C.S. Lee, and S.T. Lee

Improving Polymer Solar Cell Through Efficient Solar Energy Harvesting 199
Hsiang-Yu Chen, Zheng Xu, Gang Li, and Yang Yang

Index 237

Overview and Highlights of WOLEDs and Organic Solar Cells: From Research to Applications

Maggie Mei-Yee Chan, Chi-Hang Tao, and Vivian Wing-Wah Yam

Contents

1 Introduction 2
2 White Organic Light-Emitting Devices 3
   2.1 Advantages of WOLEDs 3
   2.2 Efficiency Characterization of WOLEDs 4
   2.3 WOLED Architectures 6
   2.4 Challenges of WOLEDs 12
   2.5 Lighting for the Future 13
3 Organic Solar Cells 14
   3.1 Basic Working Principle of Organic Photovoltaic Devices 17
   3.2 Device Characteristics of an OPV Device 19
   3.3 OPV Device Architecture 22
   3.4 Challenges of OPV Devices 30
4 Conclusions 31
References 31

Abstract Solid-state organic devices are at the vanguard of new generation of electronic components owing to their promise to be easily manufactured onto flexible substrates that potentially reduce the mass production cost for large modules. With the great efforts on improving the power efficiency that meets the realistic requirements for commercial applications, white organic light-emitting devices (WOLEDs) and organic solar cells have attracted much attention over the past two decades and are targeted as the effective ways for reducing the energy consumption and developing renewable energy in the world. Because of their great potentials to generate tremendous savings in both cost and energy usage, WOLEDs are considered as new generations of solid-state lighting sources to replace the incandescent bulbs, while organic solar cells are the most promising candidates to complement the inorganic silicon solar cells for electricity generation. Here, we

---

M.M.-Y. Chan, C.-H. Tao, and V.W.-W. Yam (✉)
Department of Chemistry, The University of Hong Kong, Pokfulam Road, Hong Kong, China
e-mail: wwyam@hku.hk

will provide a survey on the recent developments of WOLEDs and organic solar cells and their current status in these fields. Resistances and hampers to the widespread acceptances of these two areas of developments are also discussed.

1 Introduction

Energy is essential to our life, and the use of energy is increasing with human advancement and industrial development. According to the recent report on the international energy outlook (IEO) [1], the energy information administration (EIA) predicts that the world demand for energy will increase by 44% from 2006 to 2030, in which the total world energy consumption will rise from 472 quadrillion British thermal units (QBtu), or Quads, in 2006 to 678 Quads in 2030 at an average annual rate of 1.8%. Strong gross domestic product (GDP) growth and industrialization in emerging markets, especially in China and India, will drive the fast-paced growth in energy demand. Particularly, total energy demand in these emerging markets increases by 73%, compared with an increase of 15% in the industrialized nations such as the United States and Europe.

Currently, nearly all energy production comes from the burning of fossil fuels, including oil, coal, and natural gas. With the current trend of energy use, approximately 17,000 million tons of oil or 24,400 million tons of coal, will be consumed to meet the tremendous demand for energy in 2030. These increasing energy demands continue to push oil price higher and higher. The average world oil prices increased each year between 2003 and 2008. Spot prices reached US$147 per barrel in mid-July 2008. In the IEO 2009, EIA predicts that the price of oil will rise from US$61 per barrel in 2009 to US$130 per barrel in 2030 [1]. The worldwide increase in energy demand has put an ever-increasing pressure on identifying and implementing ways to save energy and to take measures to promote it. Indeed, many nations have committed on their new energy policies to improve energy efficiency. For example, Australia became the first country to announce an outright ban on incandescent bulbs starting in 2010. On 17 December 2007, President George W. Bush signed into law a landmark energy bill to begin phasing out traditional incandescent bulbs starting in 2012, to be completed in 2014. In particular, lighting accounts for a significant part of electricity consumption in the world. For instance, about 40% of all consumed energy (40.25 of 100 Quads) was used in commercial and residential buildings in the United States in 2007, and 14% of that (5.74 Quads) was used just for lighting [2]. In fact, consumers and businesses spend approximately US$58 billion a year to light their homes, offices, streets, and factories. It is clear that increasing the efficiency of lighting by a small amount has the potential to generate tremendous savings in both cost and energy use. In fact, the US Department of Energy is now anticipating that solid-state lighting in the form of white inorganic light-emitting diodes (WLEDs) and white organic light-emitting devices (WOLEDs) will decrease national energy consumption by 29% by 2025, cutting US$125 billion off the national energy bill and deferring the construction of forty 1,000 MW power plants for electricity generation [3].

Meanwhile, many nations dedicate to develop renewable energy for promoting their energy supply. On 29 April 2009, US President Barack Obama pledged to achieve its renewable objectives to make up 20% of power to be generated by wind, and for solar power to be market competitive by 2015. Renewable energy is the kind of energy generated from natural resources, including sunlight, wind, rain, tides, and geothermal heat, which can be replenished rapidly and constantly. Harvesting solar energy is an attractive option among the alternatives and presents the greatest opportunity for meeting the energy demand in a clean and renewable way. According to the US Department of Energy, the energy from sunlight strikes the earth in one hour (approximately 120,000 TW) is more than the energy we consumed in a year (~15 TW) [4]. In this chapter, we survey the recent developments on WOLEDs and organic solar cells and the current status and challenges in these fields.

2 White Organic Light-Emitting Devices

Lighting is an essential commodity of our life since the invention of incandescent light bulb by Thomas Edison in 1880. According to the Navigant 2002 annual report [5], the incandescent lamps were responsible for the 86% of the total lighting electricity consumption in the United States residential buildings, while fluorescent lamps shared only 14%. However, approximately 90% of the power consumed by an incandescent light bulb is lost as heat, rather than as visible light. With the efforts of researchers and industries, various white lighting sources, including compact fluorescent lamp (CFL), fluorescent tube, WLED, and WOLED, have been introduced. Especially, due to the significant energy saving as well as the increased lifetimes over standard incandescent lighting, WLEDs and WOLEDs have been considered as the new generation of lighting sources to replace the incandescent lamps in the near future.

2.1 Advantages of WOLEDs

Compared with the inorganic counterpart, WOLEDs have several advantages as a replacement of conventional lighting. In fact, WOLEDs were targeted toward display applications for use primarily as liquid-crystal display backlights. As their power efficiencies have surpassed those of incandescent lamps due to improvements in device architectures, and synthesis of novel materials, interest in the application of WOLED technology for solid-state lighting has been steadily increasing. A comparison of various lighting sources is provided in Table 1.

Table 1 Power efficiency (η_P), power consumption at brightness of 800 lm, chromaticity coordinates, CCT, CRI, and lifetime of various lighting sources available on the market

Type	η_P (lm W^{-1})	Power consumption (W)	CCT (K)	CRI	Lifetime (h)
Incandescent lamp	15	60	2,854	100	1,500
CFL (warm)	60	15	2,940	82	10,000
LED	90	8	3,000	80	60,000
WOLED	102[a]	–	3,900	70	–

[a]Maximum power efficiency of lighting sources reported in the literature

WOLEDs offer distinct properties to replace classical light bulbs or even fluorescent tubes. In particular, WOLEDs can be used as an ultrathin area light source, emitting diffuse light from a potentially large active area. They do not need light distribution elements, and its thickness could reach less than 1 mm that could allow the lighting to be placed directly on ceilings as a planar sheet of light. WOLEDs are also fully dimmable and can be switched on and off without any time delay. It is worth noting that being an area light source, the heat generation inside the device is not focused on a small volume as it is the case in LEDs; particularly, during standard operation, the self-heating of an OLED is limited to only a few degrees. WOLEDs are also possible to have transparent panels, which could be used as lighting elements in windows, screens, or room dividers. Most importantly, WOLEDs can be fabricated onto flexible substrates to provide new architectural design opportunities that cannot be realized with any other technology.

2.2 Efficiency Characterization of WOLEDs

WOLEDs show promise as substitutes of incandescent or fluorescent bulbs in future ambient lighting, due to favorable properties such as large area emission and potential fabrication on flexible substrates. Many smart materials and novel device configurations have been used to improve the device performance; particularly, recent demonstrations have shown continually improved WOLEDs that are now surpassing incandescent lamps in terms of efficiency and lifetime [6–8]. To turn a WOLED into a power-efficient light source, four key parameters must be addressed for evaluating the performance of WOLEDs (1) luminous power efficiency, (2) the Commission Internationale de L'Éclairage (CIE) chromaticity coordinates, (3) the correlated color temperature (CCT), and (4) the color rendering index (CRI).

Luminous power efficiency, η_P, or simply called "power efficiency", in lumen per Watt [lm W^{-1}] is the ratio of luminous power emitted in the forward direction, L_P [lm], to the total electrical power required to drive the OLED at a particular voltage, V, and current, I_{OLED}, viz.: $\eta_P = L_P/I_{OLED} V$ [9]. It is very useful in interpreting the power dissipated by a device when used in a display or lighting source. Lumen is a measure of the amount of light given off by a light source. A typical 60 W incandescent bulb gives off 840 lm, and a 100 W incandescent bulb gives off

1,750 lm. As a lighting source, the η_P of WOLEDs should be larger than that of fluorescent tubes (~60 lm W^{-1}), which is the current benchmark for novel light sources.

CIE chromaticity coordinates is the predominant, international standard for color specification and measurement. Generally, the CIE color system established in 1931 is based on the eye response of standard observers on three specific wavelengths of light in the red–green–blue (RGB) regions (700.0 nm, 546.1 nm, and 435.8 nm, respectively) [10]. Three *XYZ* tristimulus values are then derived from the relative amounts of these characteristic wavelengths present in a color. The tristimulus values *XYZ* are useful for defining a color, but the results are not easily visualized. Instead, a color space (*Y, x, y*) is commonly used. *Y* is the intensity of light (that is identical to the tristimulus value *Y*), and *x, y* are chromaticity coordinates calculated from the tristimulus values *XYZ* by

$$x = \frac{X}{X+Y+Z}, y = \frac{Y}{X+Y+Z}. \tag{1}$$

Figure 1 shows a CIE 1931 chromaticity chart for this color space. All perceived colors could be defined in this horseshoe shaped region, at which the achromatic point (0.33, 0.33) represents the "pure" white light.

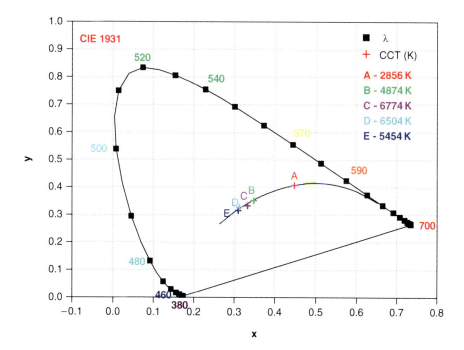

Fig. 1 CIE 1931 chromaticity chart with CRI of the common standard illuminants

CCT is the temperature of a blackbody radiator that has a color that most closely matches the emission from a nonblackbody radiator [11]. It is a measure used to describe the "whiteness" of a light source. The higher is the color temperature, the cooler the white light will be. Incandescent bulbs and CFLs are usually around 2,800 K and fall in the 2,700–3,000 K range, respectively, which are "warm" light, while a cool white light LED has a CCT of 6,500 K.

CRI of a white light source is a measure of the color shift that an object undergoes when illuminated by the light source as compared to the color of the same object when illuminated by a reference source of comparable color temperature [12, 13]. It is an international system used to rate a lamp's ability to show an object's color. The higher the CRI, based on a scale of 0–100, the truer colors will appear. Acceptable illumination sources require a CRI of higher than 80, provided that the natural sunlight has a CRI of 100.

It is worth noting that the color of two white light sources may appear identical, when viewed directly, and will therefore have the same chromaticity; however, the color of the reflected light from an object illuminated by these two sources may be significantly different, and thus the light sources will have a different CRI. For high-quality white light illumination, sources with CIE coordinates similar to that of a blackbody radiator with a CCT between 2,500 and 6,500 K, and a CRI above 80 are required. Table 2 illustrates the color quality of the common CIE standard illuminants.

2.3 WOLED Architectures

White light emission is achieved by mixing the complementary (e.g., blue and yellow or orange) colors or the three primary colors (red, green, and blue). With the continuous efforts on research activities, the η_P of WOLEDs have surpassed those of incandescent sources and are now approaching that of fluorescent tubes. Several strategies have been proposed to generate white light, including (1) doping fluorescent or phosphorescent materials into one or more light-emitting layers (EMLs) [14–23], (2) synthesis of novel metal complexes incorporating different color-emitting moieties [24, 25], (3) use of excimer or exciplex systems [26–31], (4) tandem structure [32–36], and (5) down-conversion architectures [37, 38]. Among these approaches, the incorporation of phosphorescent emitters is the most effective way for achieving WOLEDs. The use of phosphors can harvest both the singlet and

Table 2 CCT and CIE 1931 coordinates of the common CIE standard illuminants [12]

Illuminant	CCT (K)	CIE 1931
A	2,856	0.4476, 0.4075
B	4,874	0.3485, 0.3517
C	6,774	0.3101, 0.3163
D65	6,504	0.3127, 0.3291
E	5,400	0.3333, 0.3333

triplet excitons (generated at a ratio of 1:3 owing to their statistical spin population) generated by electrical injection, corresponding to a fourfold increase in internal quantum efficiency, that is the ratio of the total number of photons generated within the device to the number of electrons injected, compared to that achievable in singlet-harvesting fluorescent WOLEDs. Indeed, a high η_P of 102 lm W^{-1} has been recently realized in all phosphor-doped WOLEDs [6]. Particularly, phosphorescent emitters are doped into conductive hosts in either single emitting layer or multiple EMLs, where two or more emission centers are physically separated from each other. In general, there are three device architectures of phosphorescent WOLEDs that are often exploited. They are described as follows.

2.3.1 WOLEDs with Single Emitting Layer

Incorporation of two or three phosphorescent materials into a single host to generate white light is the simplest method for achieving WOLEDs. The key feature of this approach is the employment of one EML that comprises of a large bandgap host doped with two or more emissive phosphorescent dopants to generate white light [14–18]. A desired electroluminescence (EL) performance can be achieved just by optimizing the composition of this EML and the adjacent functional layers. This method allows the fabrication processes to be simplified as well as reduces structural heterogeneities, and produces rather stable EL spectra, suggesting its great potential for mass production and commercial realization in the future. In 2004, D'Andrade et al. [14] demonstrated the performance of WOLED that employed an EML containing three metal organic phosphors: 2 wt% iridium(III) bis(2-phenyl-quinolyl-$N,C^{2\prime}$)acetylacetonate [PQIr] providing red emission, 0.5 wt% fac-tris(2-phenylpyridine)iridium [Ir(ppy)$_3$] for green emission, and 20 wt% iridium(III) bis (4$^\prime$,6$^\prime$-difluorophenylpyridinato)tetrakis(1-pyrazolyl)borate [FIr6] for blue emission, all simultaneously codoped into a wide energy gap p-bis(triphenylsilyl)benzene (UGH2) host. Such triple-doped WOLED exhibited a maximum η_P of 42 lm W^{-1} [corresponding to an external quantum efficiency (EQE) of 12%], and CIE coordinates of (0.43, 0.45) with CRI of 80. Eom et al. [16] further improved the EL efficiency of these WOLEDs by utilizing dual triple-doped emissive layers, in which three phosphorescent dopants [PQIr, Ir(ppy)$_3$, and FIr6] were codoped into two adjacent wide bandgap hosts, namely, $N,N^\prime$-dicarbazolyl-3,5-benzene (mCP) and UGH2. Figures 2 and 3 show the chemical structures of the phosphors and the host materials employed for WOLEDs, respectively. This double EML structure broadened the exciton recombination zones, resulting in a significant improvement in η_P of 26 lm W^{-1}, at a luminance of 1,000 cd m^{-2}, doubling that of triple-doped WOLEDs reported by D'Andrade et al. ($\eta_P = 11$ lm W^{-1}) [14]. With balanced emission from the three emitting layers, white light emission with CRI of 79 and CIE coordinates of (0.37, 0.40) could be achieved.

WOLEDs can also be prepared by the combination of two complementary colors (blue and yellow or orange). Wang et al. [17] recently reported a highly efficient WOLED by incorporating two phosphorescent dyes, namely, iridium(III)

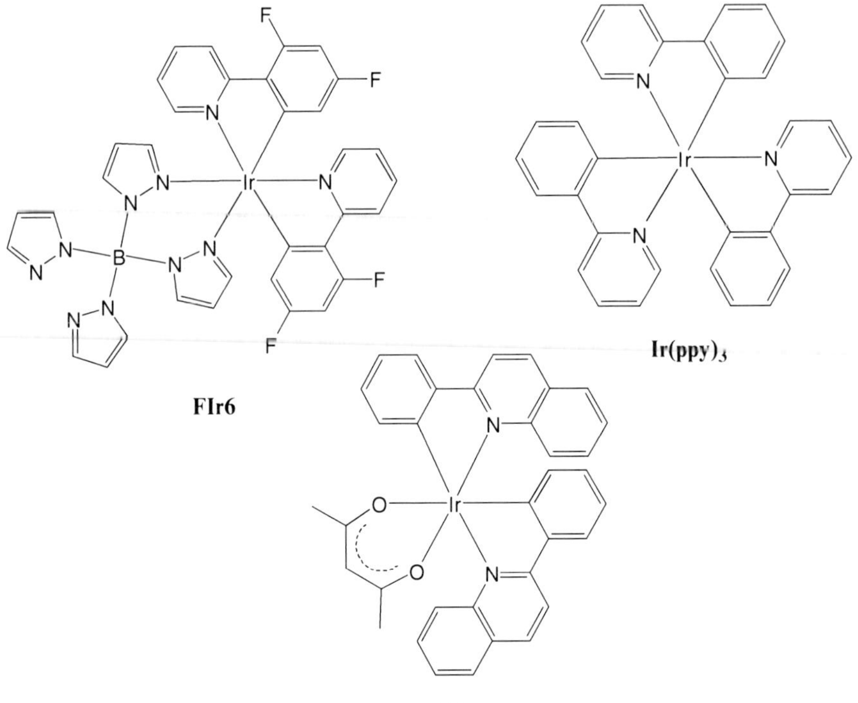

Fig. 2 Chemical structures of phosphors used for WOLEDs

[bis(4,6-difluorophenyl)-pyridinato-N,C^{2}]picolinate [FIrpic] doped into mCP for blue emission and bis(2-(9,9-diethyl-9H-fluoren-2-yl)-1-phenyl-1H-benzoimidazol-N,C^{3})iridium (acetylacetonate) [(fbi)$_2$Ir(acac)] doped into 4,4′,4″-tri(N-carbazolyl)triphenylamine (TCTA) for orange emission, into a single-energy well-like emissive layer. Figure 4 depicts the chemical structures of [(fbi)$_2$Ir(acac)] and TCTA. Such device achieved a peak forward-viewing η_P of 42.5 lm W^{-1}, corresponding to EQE of 19.3%, a current efficiency (or sometimes called "luminous efficiency") of 52.8 cd A^{-1}, and excellent color stability with CIE coordinates of (0.30, 0.37) at luminance of 1,000 cd m^{-2}. Particularly, the key feature for extremely high η_P and EQE is related to the careful manipulation of two exciton-formation modes, namely host–guest energy transfer for the blue dopant and direct exciton formation for the orange dopant within an energetic, well-like, single emissive region. This unique strategy created two parallel pathways to channel the overall excitons to both dopants within the EML, leading to an improved charge balance and further reduction of the unfavorable energy losses. Hence, an extremely high efficiency WOLED with nearly 100% internal quantum efficiency could be realized.

However, in most conventional single-EML WOLEDs, it is quite difficult to obtain a common host material with a wide bandgap and large triplet energy

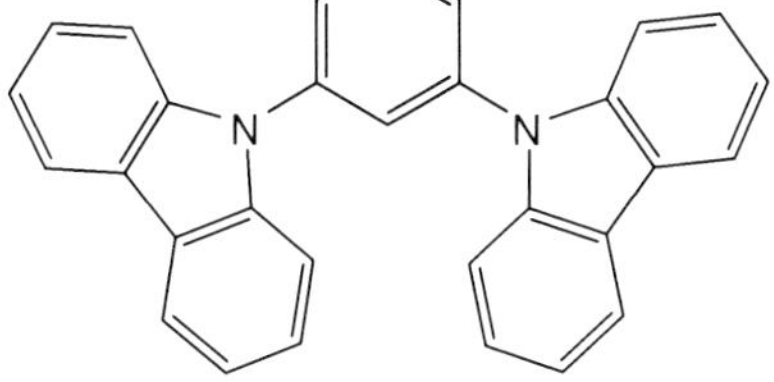

mCP

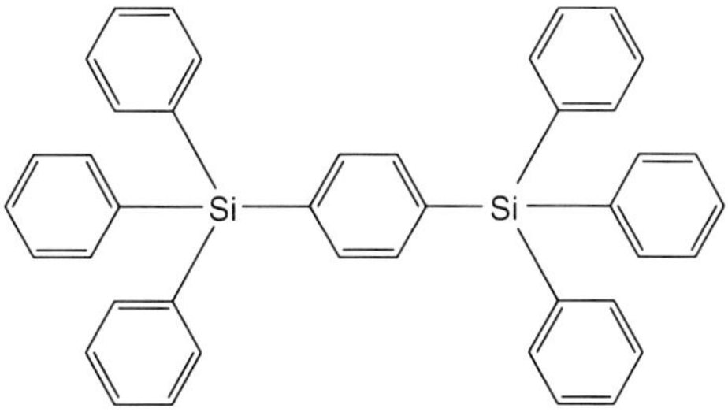

UGH2

Fig. 3 Chemical structures of host materials used for WOLEDs

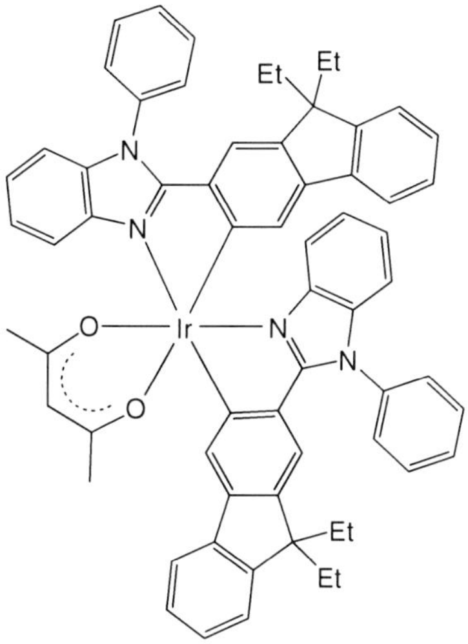

$(fbi)_2Ir(acac)$

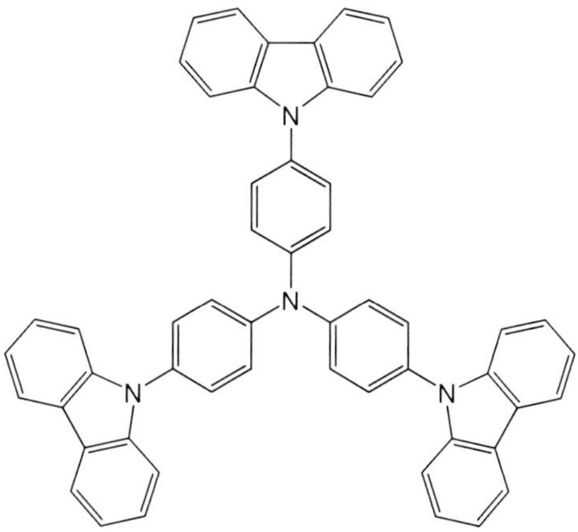

TCTA

Fig. 4 Chemical structures of $[(fbi)_2Ir(acac)]$ and TCTA

(>2.8 eV) to efficiently transfer energy from the host to all the three dopants [15, 18]. In addition, sequential energy transfer initially from the short-wavelength dopant always dominates the main emission mechanism. Indeed, the majority of excitons are generated by Dexter energy transfer from the host or trapping on the blue-emitting molecules, and then subsequently transferred to the green and red dopants [19, 20]. Hence, the EQE of the entire device is limited by the less-efficient blue species. Furthermore, host materials that have wider bandgaps will act as energy barriers for the transport of carriers from nearby hole or electron-transporting layer to EMLs, which consequently decrease the probability of carrier recombination. Particularly, WOLEDs with single EML usually have higher operating voltage [17].

2.3.2 Phosphorescent WOLEDs

To date, the most impressive characteristics of WOLEDs reported have been achieved in all-phosphor-doped devices, in which three phosphorescent dopants (red, green, and blue) are intentionally doped into three separate EMLs [20, 21]. The key feature of this WOLED is the positioning of blue phosphor within the emission layer and its combination with a carefully chosen host material: energetically, (1) the triplet energy of the blue emitter material is in resonance with its host, so that the blue phosphorescence is not accompanied by internal triplet energy relaxation before emission and (2) triplet energies of all materials, that is, the host materials and the adjacent hole- and electron-transporting materials, should have higher triplet energies than that of the phosphorescent emitters. This can essentially define the exciton distribution within the multilayer emission layer. With good knowledge on device engineering, multiple EMLs effectively broaden the exciton recombination zone and reduce the pile-up of excitons at the EML, thereby enhancing EL efficiency and reducing the efficiency roll-off. In particular, triplet–triplet annihilation is the main mechanism for unusual strong efficiency roll-off at high current densities and luminance [22]. It is reported that the degree of triplet–triplet annihilation is proportional to the square of the triplet exciton density, and a narrow emission zone has been found to have a negative effect on triplet–triplet annihilation due to its high triplet exciton density [22]. Maximizing the recombination zone thus effectively suppresses the triplet–triplet annihilation and reduces the efficiency roll-off.

Sun and Forrest [21] had successfully demonstrated a highly efficient WOLED by employing three adjacent phosphorescent EMLs. The phosphorescent dopants, namely, [PQIr] for red, [Ir(ppy)$_3$] for green, and [FIr6] for blue emissions, were each doped in separate hosts that form a stepped progression of highest occupied and lowest unoccupied molecular orbitals. This structure effectively broadened exciton generation region, as well as allowed for separate optimization of the three dopant–host material combinations across the three hosts. Such 3-EML WOLED exhibited high peak forward-viewing η_P and EQE of 32 lm W^{-1} and 16%, respectively, and good CIE coordinates of (0.37, 0.41) with CRI of 81. Very recently, WOLED with fluorescent tube efficiency had been realized, as reported by Reineke et al. [6]. This three-EML device combined a novel concept for energy-efficient photon generation with an improved out-coupling, in which blue-emitting layer was surrounded by red and green sublayers of the emission layer to harvest unused excitons. By combining a carefully chosen emitter layer with high refractive index substrates, and using a periodic out-coupling structure, an extremely high η_P of 90 lm W^{-1} at luminance of 1,000 cd m^{-2} could be achieved. This efficiency has the potential to be raised to 124 lm W^{-1} (corresponding to EQE of 46%) if the light out-coupling can be further improved.

As the η_P of WOLEDs employing multiple EMLs approaches fluorescent tube efficiency, it requires a careful control of the location of exciton recombination and the energy transfer/exciton diffusion between or within layers, via the changing of layer thickness, doping concentration, and charge blocking layers, in order to obtain a balanced white light emission. In addition, as one of the three primary colors, the maximum EQE of blue OLEDs are comparatively lower than those reported for green and red OLEDs [19]. While suitable molecules have been found for the green and red, stable and efficient blue phosphorescent emitters are still a challenge, and this inevitably limits the efficiency of WOLEDs with multiple EMLs.

2.3.3 Hybrid WOLEDs

For power-efficient phosphorescent WOLEDs, an additional challenge is that high energy phosphors demand host materials with even higher triplet energies to confine the exciton to the emitter. This problem can be eliminated by introducing a novel device concept that exploits a blue fluorescent molecule with high triplet energy in exchange for a phosphorescent dopant, in combination with green and red phosphor dopants [8, 19, 20, 22], to yield high η_P and stable color balance, while maintaining the potential for unity internal quantum efficiency. In short, this structure takes advantage to harness all electrically generated high-energy singlet excitons for blue emission, and phosphorescent dopants to harvest the remainder of lower energy triplet excitons for green and red emission.

In 2006, Sun et al. [19] illustrated the effectiveness of this WOLED architecture by doping the middle region of the EML with both the green [Ir(ppy)$_3$] and red [PQIr] phosphorescent dopants sandwiched between two blue fluorescent-doped EMLs. The key feature of this WOLED structure was that the phosphor-doped region was separated from the exciton-formation zones by spacers of undoped host material. The triplet excitons then diffused efficiently to the central region, where they transferred to the lower energy green or red phosphor dopants, again by a nearly resonant process to the green dopant triplet manifold, and with some energy lost to the red triplet. Diffusion of singlet excitons to the phosphorescent dopants was negligible due to their intrinsically short diffusion lengths. The optimized device exhibited a high η_P and total EQE of 37.6 lm W^{-1} and 18.7%, respectively. In addition, the CIE coordinates had a negligible shift from (0.40, 0.41) at low current densities to (0.38, 0.40) at 100 mA cm^{-2}. It is also worth noting that the optimized WOLED had a less-pronounced efficiency roll-off at high current densities, as compared to previous all-phosphor, high-efficient WOLEDs.

Schwartz et al. [22] modified this WOLED architecture by employing a fluorescent blue emitter, N,N'-di-1-napthalenyl-N,N'-diphenyl-[1,1':4',1'':4'',1'''-quaterphenyl]-4,4'''-diamine (4P-NPD) (as shown in Fig. 5), with very high triplet energy of 2.3 eV and high photoluminescence (PL) quantum yield of 92%, rendering it possible to harvest its triplet excitons by letting them diffuse to an orange phosphorescent iridium complex. Such WOLED exhibited a total η_P of 57.6 lm W^{-1} at a brightness of 100 cd m^{-2} (corresponding to an EQE of 20.3%). Later, they demonstrated an improved device structure by directly doping the orange phosphorescent material into the blue fluorescent EML with low concentration (0.2 wt%) to decouple the strong efficiency roll-off [8]. The modified WOLED

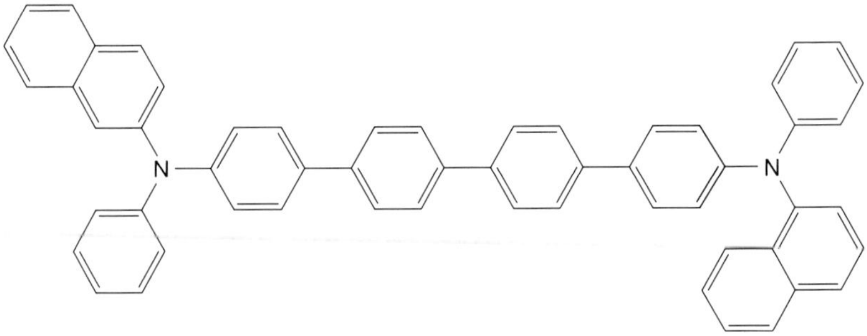

Fig. 5 Chemical structure of 4P-NPD

yielded a total η_P of 49.3 lm W^{-1} and total EQE of 24.1%, measured in an integrating sphere, and CIE coordinates of (0.49, 0.41) with CRI of 82.

A key prerequisite for this hybrid WOLED concept is choosing a suitable blue fluorescent-emitting material that should have a high triplet energy as well as high PL quantum yield. However, up until now, there is only limited choice on the blue fluorescent emitter. Particularly, triplet excitons, which are either directly generated on or transferred by Dexter transfer to the blue emitter, are being lost [22]. This is a result of the excitons' inability to leave the blue emitter because of its low-lying triplet state. Hence, rather low efficiencies can only be achieved for most of the hybrid WOLEDs at an illumination relevant brightness, say, 1,000 cd m^{-2} [8].

2.4 Challenges of WOLEDs

As the development of novel organic materials and smart device design further improves the performance of WOLEDs, the major challenge for the commercialization of WOLED as solid-state lighting is the operational lifetime. Typically, fluorescence bulbs have lifetimes of ~10,000 h. However, the operational lifetimes of the reported blue phosphorescence devices are significantly short; for example, the luminance of the FIrpic-based blue phosphorescent device drops to 50% of its initial value within 1–2 h [6]. To compete as a light source in the general illumination, market will require a product life of tens of thousands of hours. End of life will be determined when the original luminance intensity of the device decreases to 70% of the original value, versus 50% as often referenced for display applications. As the product ages, the color is not expected to change. In other words, the relative intensity of the respective color sources should remain constant over the life of the product.

Another major challenge facing this WOLED technology for general illumination purposes is reducing the cost per lumen. As lighting is not a new market, WOLEDs must be cheaper than the current technology in order to be adopted. The decorative lighting market has the least stringent requirements and requires a brightness of 50–500 cd m^{-2} at an efficiency of >10 lm W^{-1} for an operational lifetime of at least 10,000 h. The market for general illumination requires much larger panels at high brightness (5,000 cd m^{-2}) and very high efficiencies of >50 lm W^{-1}. If the cost of an 800 lm WOLED comes down even to US$3.00 per megalumen-hour, it makes economic sense to be widely adopted.

2.5 Lighting for the Future

Through continuing efforts on improving luminance efficiency, color gamut, and device reliability, WOLEDs have a great potential to replace or complement other lighting technologies and becomes the light source of choice for the future. According to the recent NanoMarkets report "OLED Lighting Markets" [39], there are great commercial opportunities for OLED lighting used for backlighting, general illumination, specialty/architectural lighting, vehicular lighting, signage, and niche applications. Meanwhile, the US Department of Energy expects OLED lighting to reach a high η_P of 150 lm W^{-1} in 2012 rather than 2014 as previously forecasted [40]. NanoMarkets believe that these and other improvements in OLEDs will drive the general illumination market to US$2.3 billion in revenues by 2015. In particular, the OLED backlighting market will reach US$1.1 billion by 2015, while OLED architectural and specialist industrial lighting will reach US$1.9 billion.

With the advent of commercial interest in WOLED, many anticipative applications have been demonstrated. For example, on the Plastic Electronics Asia 2009, Fraunhofer Institute for Photonic Microsystems IPMS demonstrated a steering wheel with an integrated OLED signage device for automotive applications (Fig. 6) [41]. This OLED device was made within the German joint development

Fig. 6 Steering wheel with an integrated OLED signage device demonstrated by the Optrex Europe GmbH, Fraunhofer Institute for Photonic Microsystems IPMS, and G. Pollmann GmbH under the German joint development project CARO (Car OLED). Reproduced with permission from [41]. Copyright 2009, Optrex Europe GmbH

project CARO (Car OLED) among Optrex Europe GmbH, Fraunhofer Institute for Photonic Microsystems (Fraunhofer IPMS), and the styling-studio G. Pollmann GmbH. Meanwhile, OSRAM Opto Semiconductors GmbH had developed a transparent WOLED with outstanding performance (Fig. 7) [42]. The WOLED prototype lighted up an area of nearly 90 cm^2, with η_P of more than 20 lm W^{-1} at a brightness of 1,000 cd m^{-2} and CIE coordinates of (0.396, 0.404). This WOLED was transparent whether it is powered on or off, and its transparency was currently rated at 55%. This opens up possible applications such as partitions that are almost invisible by day and then provides a pleasant diffused light at night. In March 2008, GE Global Research and GE Consumer and Industrial successfully demonstrated "the world's first roll-to-roll manufactured OLED lighting devices," as shown in Fig. 8 [43].

In April 2008, OSRAM Opto Semiconductors GmbH presented the first WOLED product "Early Future" at the Light & Building Fair in Frankfurt, Germany [42]. This WOLED table lamp is designed by a renowned lighting designer Ingo Maurer and is being produced as a limited edition. It works with ten WOLED tiles, each measures at 132 × 33 mm, at luminance of 1,000 cd m^{-2} and η_P of 20 lm W^{-1}, as shown in Fig. 9. Mercedes Benz recently announced a new 2010 Mercedes Benz E-class equipped with an elegant control panel that uses a WOLED as the display (Fig. 10) [44]. These revolutionary application marks the emergence of a new technology of lighting.

3 Organic Solar Cells

As mentioned in Sect. 1, the global demand for energy is continually expanding, that the total world energy use will increase at an average annual rate of 1.8% and reaches 678 Quads (~18.49 TW) in 2030. Currently, nearly all energy production comes from the burning of fossil fuels, such as oil, coal, and natural gas. Through the incremental improvement on the renewable energy technologies, including hydroelectric, geothermal, tidal, wind, and biomass, harnessing solar energy is an attractive option among the different carbon-free alternatives. Table 3 summarizes the energy supply from the renewable energy technologies. Approximately 89,000 TW of solar energy strikes the earth's surface each year, and the capture of a fraction of solar energy can supply all of our energy needs in 1 year, that is, the world energy demand in 2008 is equal to 15 TW. As solar energy offers great potential for the supply of all our energy demand, solar cell technology has experienced rapid growth in recent years as the costs have improved. According to the recent research report "World Solar Cell Market: Key Research Findings 2009" issued by Yano Research Institute, the global solar cell production had been increased by 69% from 3.8 GW in 2007 to 6.5 GW in 2008 [45]. Meanwhile, many companies are planning to enter the solar cell business by building their production line. For instance, in 2008, LG Electronics, a global leader and technology innovator in consumer electronics, decided to convert its plasma panel-manufacturing

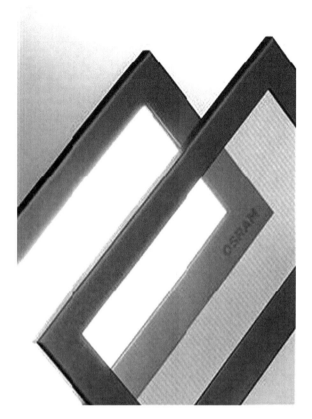

Fig. 7 Transparent WOLED prototype from OSRAM Opto Semiconductors GmbH. Reproduced with permission from [42]. Copyright 2009, OSRAM Opto Semiconductors GmbH

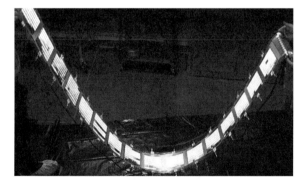

Fig. 8 The world's first roll-to-roll manufactured OLED lighting devices from GE Global Research and GE Consumer & Industrial. Reproduced with permission from [43]. Copyright 2009, General Electric Company

Fig. 9 OSRAM Early Future OLED lamp designed by a renowned designer Ingo Maurer. Reproduced with permission from [42]. Copyright 2009, OSRAM Opto Semiconductors GmbH

Fig. 10 The 2010 Mercedes Benz E-Class control panel. Reproduced with permission from [44]. Copyright 2009, Dailmer AG

Table 3 Energy supply from the renewable energy resources in 1 year

Resources	Energy supply (TW)
Hydropower	0.5
Geothermal	12
Tidal	2
Wind	5
Solar	89,000
Biomass	1.7

line in Gumi, Korea, into solar cell production lines and planned to manufacture crystalline silicon solar cells and modules with a capacity of 120 MW each [46].

Solar cells generate power by converting photons from the sun into electricity through a mechanism called "the photovoltaic effect." The term "photo" means light and "voltaic" means electricity. A solar cell is primarily constructed of inorganic semiconductor that generates electricity when sunlight falls on it. Nowadays, most of the solar cells are silicon based, in which their power conversion efficiencies (i.e., the efficiency to convert incident solar power to useful electric power) can achieve up to 23.2%, reported by a research team from Eindhoven University of Technology together with the Fraunhofer Institute in Germany [47]. The US Department of Energy's National Renewable Energy Laboratory (NREL) has recently developed a triple-junction solar cell that can convert 40.8% of sunlight into electricity under concentrated light of 326 suns [48]. Meanwhile, the University of Delaware has broken the world record for solar cell to 42.8% by using an optical concentrator to split light into components, i.e., high, medium, and low energy light, and direct it to several different materials which can then extract electrons out of its photons [49]. With the great effort on improving the efficiency of solar cells, their costs have been rapidly reduced. In 1956, silicon cells cost approximately US$300 per Watt of electricity produced. Today, the solar cells have dropped in cost to approximately US$1.3 per Watt [50]. However, this cost is still too high to be competitive against relatively plentiful and cheap fossil fuels. In order for large-scale adaptation of this technology, the cost of solar cells needs to be dropped by at least a factor of 10 in order to be on par with current means of electricity generation. There is a prevailing need for the development of new materials and concepts for photovoltaic energy conversion that could potentially reduce the manufacturing cost of solar cells.

3.1 Basic Working Principle of Organic Photovoltaic Devices

The development of dye-sensitized solar cell and organic photovoltaic (OPV) devices, so-called "third and fourth generation" solar cells, has been envisaged as a possible route. In this chapter, the field of OPV devices will be briefly introduced. OPV devices provide many foreseeable advantages, including simple fabrication process, light weight, flexibility, and solution processability for large area production. These, together with the virtually infinite variety of molecules to choose from and design that show very strong optical absorption with tunable absorption and electrical properties, would make them very attractive. OPV devices generally consist of two photoactive organic materials (so-called "donor–acceptor" system) sandwiched between two conducting electrodes. Under light illumination, excitons are generated in the photoactive layers and some would dissociate into electrons and holes at the donor–acceptor heterointerfaces. Alternatively, some excitons may diffuse to the electrode and decay nonradiatively via quenching and releasing energy in the form of a phonon as heat. These charge carriers are subsequently collected at the respective electrodes, contributing to the photocurrent [51–54]. It is worth noting that the process of photocurrent generation by an OPV device is analogous to the basic working principle of OLED. Particularly, OLED is a device that converts electrical energy into light, while OPV is a device that converts the energy of sunlight directly into electricity. Figure 11 illustrates the schematic diagram of the basic working principle of an OPV device. The process of converting light into electric current in an organic OPV device is accomplished by four consecutive steps (1) Light absorption leading to the formation of an excited electron–hole pair (exciton); (2) exciton diffusion to the donor–acceptor heterointerfaces, where (3) the exciton dissociation occurs into a free electron and hole by charge transfer; (4) finally the charge carriers are collected at the anode (holes) and cathode (electrons), providing current to a load in the external circuit. In short, the external quantum efficiency (η_{EQE}) of an OPV device is defined as the ratio of the number of electron–hole pairs collected at the electrodes to the number of incident photons [52, 53], viz.,

$$\eta_{EQE} = \eta_A \eta_{ED} \eta_{CT} \eta_{CC}, \qquad (2)$$

where η_A, η_{ED}, η_{CT}, and η_{CC} are the absorption, exciton diffusion, charge transfer, and charge carrier collection efficiencies, respectively. Absorption efficiency, η_A, is the fraction of photons absorbed, and it can simply be increased by using thicker

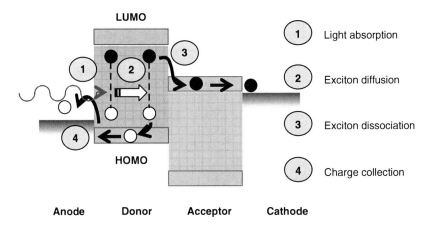

Fig. 11 Schematic diagram of the basic working principle of an OPV device

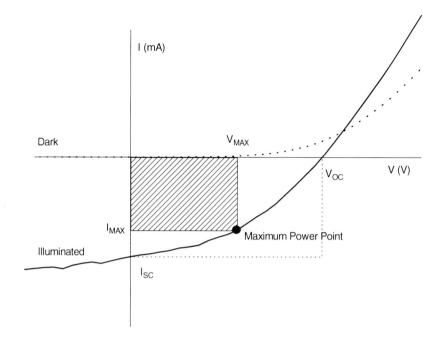

Fig. 12 Typical current–voltage curves of an organic solar cell in the dark and under illumination. Important device parameters are also shown, including short-circuit current (I_{SC}), open-circuit voltage (V_{OC}), and the points of current and voltage (I_{MAX} and V_{MAX}, respectively) that correspond to maximum power output

photoactive layers. However, exciton diffusion length invariably limits the thickness of the photoactive layer. Only photons absorbed within the exciton diffusion length from the donor–acceptor interface can survive the transport to the interface to subsequently dissociate into free charge carriers with a high yield, defined as the exciton diffusion efficiency, η_{ED}. The inherent trade-off between the η_A and the η_{ED} poses an upper limit of η_{EQE}. Fortunately, most organic materials possess high absorption coefficients of larger than 10^5 cm^{-1}, providing the possibility for keeping the layer thickness as thin as possible (~exciton diffusion length) with highly absorptive properties.

3.2 Device Characteristics of an OPV Device

In order to maximize the device performance, it is crucial to understand the parameters of an OPV device that form the rationale for the design of new organic materials or device architectures. A graph of current–voltage (I–V) characteristics is a common way to illustrate the photovoltaic response of an OPV device. Figure 12 depicts the typical I–V curves of an OPV device. In the dark, there is almost no current flowing through the devices. When the forward bias voltage becomes larger than the open-circuit voltage, current starts to flow. Under illumination, the current flows in a direction opposite to that of the injected currents, that is, I–V curve shifts downward. From the I–V curves, four device parameters can also be determined.

3.2.1 Short-Circuit Current

Short-circuit current (I_{SC}) is the current under zero-applied bias voltage. It is the current that flows through an illuminated solar cell when there is no external resistance, that is, when the electrodes are simply connected or short circuited. I_{SC} is the maximum current that a device is able to produce, and it is strongly dependent on the absorption efficiency and exciton diffusion length in the photoactive materials. Particularly, it is linearly proportional to the η_{EQE} [54].

3.2.2 Open-Circuit Voltage

Open-circuit voltage (V_{OC}) is the voltage at which the photogenerated current is balanced to zero (so-called flat band condition). It is the maximum possible voltage across an OPV device. V_{OC} is predominantly determined by the energy difference between the highest occupied molecular orbital (HOMO) of the donor and the lowest unoccupied molecular orbital (LUMO) of the acceptor [55, 56], that is, $qV_{OC} = E_{HOMO}$ (donor)–SE_{LUMO} (acceptor), where q is the electronic charge, E_{HOMO} and E_{LUMO} are the energy levels of the photoactive materials, and S is the slope of the linear fit.

3.2.3 Fill Factor

Fill factor (FF) is the ratio of the maximum power output of an OPV device to its theoretical power output if both current and voltage were at their maxima, I_{SC} and V_{OC}, respectively [57]. It is used to characterize the shape of the I–V curves in the power-generating fourth quadrant.

$$FF = P_{MAX}/I_{SC} \times V_{OC} = I_{MAX} \times V_{MAX}/I_{SC} \times V_{OC}. \tag{3}$$

For a high FF, the shunt resistance should be very large to prevent leakage currents, and the series resistance should be very small to get a sharp rise in the forward current.

3.2.4 Power Conversion Efficiency

Power conversion efficiency (PCE) is the key quantity used to characterize the cell performance. It is a standard industry metric for solar cell performance, and it measures the ratio of the electrical power produced by the cell per unit area in watts, divided by the watts of incident light under certain specified conditions called "standard test conditions" [50]. In other words, it is the ratio of the amount of

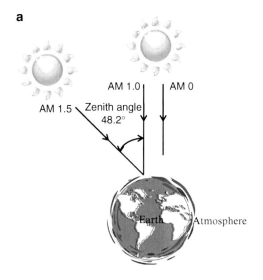

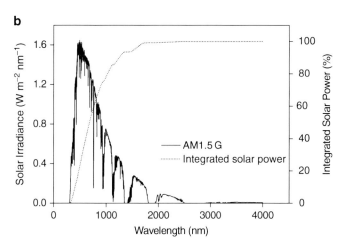

Fig. 13 (a) Schematic diagram of different AM sunlight spectra. (b) Spectral irradiance (*solid line*) for AM1.5G sunlight and the integrated solar power

optical power produced by the OPV device relative to the power available in the incident solar radiation (P_{IN}), given as

$$\text{PCE} = P_{MAX}/P_{IN} = I_{SC} \times V_{OC} \times FF/P_{IN}(\%). \quad (4)$$

Standard test conditions represent a set of conditions under which the solar cell can be evaluated and compared, including (1) irradiance intensity of 1,000 W m^{-2}, (2) Air Mass (AM) 1.5 solar reference spectrum, and (3) the solar cell temperature during measurement of 25°C.

AM is a measure of how much sunlight traveled through the atmosphere to reach the earth's surface. This is usually denoted as "AM(x)", where x is the inverse of the cosine of the zenith angle of the sun [58]. A typical value for solar cell measurements is AM1.5, which means that the sun is at an angle of 48°. Figure 13a shows the schematic illustration of different AM sunlight spectra. Figure 13b depicts the spectral irradiance for AM1.5 global sunlight and the integrated solar power. The solar spectrum includes invisible ultraviolet (UV) light, the visible spectrum of colors, and the invisible infrared (IR) spectrum. Solar radiation includes wavelengths as short as 300 nm and as long as 4,045 nm. The amount of incoming photons across the UV, visible, and IR spectra is about 3%, 45%, and 52%, respectively. It is worth noting that harvesting as large a portion of the solar spectrum is important to increase I_{SC}, particularly for the red and near-infrared spectral regions. For example, absorbing light at $\lambda \leq 1.1$ μm (that is the optical bandgap of silicon) can collect up to 77% of the incident power. Using organic material with a low optical bandgap can effectively increase the I_{SC}.

3.2.5 Incident Photon to Current Efficiency

Incident photon to current efficiency (IPCE) is the efficiency of an OPV device as a function of the energy or wavelength of the incident radiation. Under monochromatic lighting at a particular wavelength, λ, it is defined as the number of photo-generated charge carriers contributing to the photocurrent per incident photon. It is simply the number of electrons measured under short-circuit current conditions, no applied bias, divided by the number of incident photons [58]. IPCE is defined as

$$\text{IPCE} = 1.24 I_{SC}/(G \times \lambda)\,(\%), \tag{5}$$

where G is the light density (in W cm^{-2}).

3.3 OPV Device Architecture

Nowadays, most of the OPV devices consist of two photoactive materials to form donor–acceptor heterojunction. Indeed, the photovoltaic effect in organic materials was first observed in single layers of organic materials, deposited between two metal electrodes of different work functions, to produce a rectifying device. These devices produced rather low PCE of only 0.01% [54]. Particularly, the reasons for the poor performance may be attributed to the low exciton dissociation rate. In single-layer devices, only the excitons that can remain intact long enough to reach the electrode can dissociate into free charge carriers. An estimation based on Onsager model for charge–pair dissociation revealed that exciton dissociation in the bulk of an organic material requires applied electric fields in excess of 10^6 V cm^{-1} to overcome the binding energy of an electron–hole pair and to have a considerable dissociation probability [59]. The breakthrough that led to an exponential growth of this OPV field was achieved by Tang in 1985 [60]. He introduced the concept of bilayer heterojunction, in which two organic layers with specific electron- or hole-transporting properties were sandwiched between the indium tin oxide (ITO)-coated glass as anode and a Ag cathode. Using a copper phthalocyanine (CuPc) as the donor and a perylene tetracarboxylic derivative as acceptor, a remarkably high PCE of 0.95% was recorded under simulated AM2 illumination with light intensity of 75 mW cm^{-2}, an order of magnitude higher than that for single-layer OPV devices. Figure 14 illustrates a schematic diagram of a bilayer OPV device and the commonly used donor and acceptor materials, including CuPc [61–65], buckminsterfullerene (C_{60}) [61–69], 3,4,9,10-perylenetetracarboxylic bis-benzimidazole (PTCBI) [60, 64, 65], pentacene [68, 69], [6,6]-phenyl-C_{61}-butyric acid methyl ester (PCBM) [70–73], [6,6]-phenyl-C_{71}-butyric acid methyl ester ($PC_{70}BM$) [74, 75], and poly(3-hexylthiophene) (P3HT) [70, 72, 73]. The performance improvement is mainly attributed to the sharp increase in the exciton dissociation efficiency, in which the discontinuous HOMO and LUMO levels at the donor–acceptor interface can efficiently separate the bound electron–hole pairs into free charge carriers. In addition, this bilayer device architecture can also enhance the spectral coverage within the solar spectrum due to the use of two photoactive materials. The use of separated charge transporting materials to form the donor–acceptor junction can further assist the transport of charge carriers to their respective electrodes, enhancing the charge collection efficiency. For bilayer OPV devices, efficient exciton dissociation at the donor–acceptor heterojunction takes place, provided the energy level offset for the HOMO or LUMO (ΔE_{HOMO} or ΔE_{LUMO}, respectively) should be larger than the exciton binding energy (E_B) in photoactive materials [76]. In particular, ΔE_{HOMO} (ΔE_{LUMO}) should be larger than the acceptor (donor) exciton binding energy for efficient exciton dissociation in the acceptor (donor) [76]. E_B is the minimum energy needed to overcome the Coulomb electric force to separate the exciton into free electron and hole [76], and is given by the energy difference between the transport gap and optical gap of organic material, that is, $E_B = E_{TRANS} - E_{OPT}$ [76, 77]. As shown in Fig. 15, transport gap is the energy difference between the HOMO and LUMO of the organic material while optical gap is determined from the low-energy absorption edge. For most organic materials, the E_B is within 0.2–2 eV [77].

Since then, many smart device configurations and high-performance materials have recently been developed to improve the photovoltaic responses. Encouraging progress with PCE up to 6.1% has been recently reported [74]. Here, we review various architectures that have been often exploited to optimize the device performance, including the introduction of exciton blocking layer (EBL), optical spacer, bulk heterojunction, low bandgap photoactive materials, and tandem structures.

3.3.1 Exciton Blocking Layer

As metal cathode is deposited onto the photoactive materials, the metal clusters may diffuse into the organic materials, creating a highly structured interface and

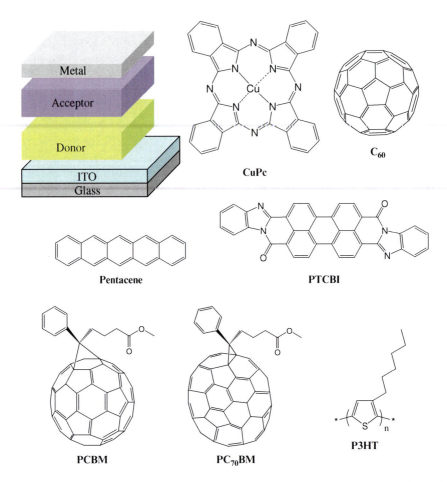

Fig. 14 Typical device architecture of an organic solar cell and the commonly used small molecular and polymeric materials

rendering interface induced exciton recombination more likely. Particularly, defect states within the organic layer are induced which quench the excitons as well as trap charge carriers, leading to a high contact resistance at the acceptor–metal interface. This definitely limits the internal quantum efficiency of OPV devices. In order to solve this problem, Peumans and Forrest introduced bathocuproine (BCP) as an EBL inserted in between the acceptor layer and metal cathode [61]. With its large ionization potential (I_P = 6.4 eV) and large energy gap (3.5 eV), BCP acts as an effective EBL which blocks the diffusion of excitons to the cathode. This confinement of excitons significantly reduces the exciton quenching at the cathode, so that more excitons can contribute to the generation of electricity, and thus leads to efficiency enhancement. Particularly, for standard bilayer CuPc/C_{60} OPV devices, a tenfold increase in PCE (~1.5%) can be achieved for devices with BCP as EBL, as compared to that without EBL layer (~0.017%) [63].

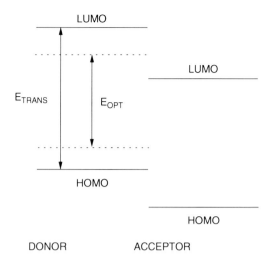

Fig. 15 Schematic energy level diagram of a donor–acceptor interface

However, the organic material employed must fulfill the basic requirements for an efficient EBL. The film needs to have a high electron affinity and a high I_P, such that it allows efficient electron injection while blocking exciton diffusion [52, 61]. Furthermore, it needs to have a low optical absorption coefficient and a high electrical conductivity, such that a good device performance can be achieved even when a large variety of optimum optical thicknesses are required for different OPV devices. This definitely limits the choice of materials that can be employed. More importantly, efficient electron transport in the EBL occurs mainly through defect states induced during the metal cathode deposition. If the EBL is too thick, the charge collection efficiency will be reduced as the cell resistance increases. Particularly, the thickness of EBL is around 5–10 nm, depending on the depth of metal atoms penetrating into the organic material.

3.3.2 Optical Spacer

Due to the high reflectivity of metal cathode, optical interference between the incident (from the transparent anode side) and back-reflected light occurs, in which the intensity of the light is almost zero at the metallic cathode [70]. This leads to a relatively large fraction of the photoactive layer positioned in a dead-zone, significantly reducing the photogeneration of charge carriers. In order to overcome this problem, a concept of optical spacer that spatially redistributes the light intensity inside the device is introduced. An optical spacer is a highly transparent layer inserted in between the acceptor layer and the cathode. It is used for modulating the intensity of the incident light, such that intensity maximum would be close to the donor–acceptor interface where absorption and charge dissociation are efficient, as shown in Fig. 16. Depending on the optical properties

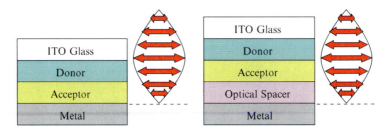

Fig. 16 Schematic diagram of the spatial distribution of the optical electric field strength inside the devices with a structure of ITO/active layer/metal and ITO/optical spacer/active layer/metal

(optical absorption spectra and refractive index) of the active organic materials used in the OPV devices, different thicknesses of an optical spacer (typically 10–30 nm) should be used to achieve the optimum optical design [70, 78–80]. Basically, the films as optical spacer must be a good electron-transporting material and is highly transparent within the solar spectrum. Furthermore, it needs to have an excellent energy level matching with the LUMO of the acceptor layer as well as the Fermi level of the metal cathode for electron injection [70, 78–80].

Recently, Kim et al. [70] demonstrated the effectiveness of this concept by using a solution-based sol–gel process to prepare a titanium oxide (TiO_x) as an optical spacer inserted in between the polymeric matrix and the Al cathode. It is well known that TiO_x is a good electron acceptor and electron-transporting material, as confirmed by its use in dye-sensitized solar cells. As expected, this amorphous TiO_x layer exhibited a high electron mobility of 1.7×10^{-4} cm^2 V^{-1} s^{-1}, and a relatively large optical bandgap of 3.7 eV. More importantly, TiO_x had a good energy level matching with the LUMO of polymeric P3HT:PCBM matrix as well as the metal cathode, satisfying the electronic structure requirements of an optical spacer. By employing TiO_x as optical spacer, the polymeric OPV devices demonstrated substantially improved photovoltaic responses. Under AM1.5 illumination with light intensity of 90 mA cm^{-2}, such device exhibited a high I_{SC} of 11.1 mA cm^{-2}, a V_{OC} of 0.61 V, a FF of 0.66, and a PCE of 5.0%. These values were significantly better than those ($I_{SC} = 7.5$ mA cm^{-2}, $V_{OC} = 0.51$ V, FF = 0.54, and PCE of 2.3%) of devices without an optical spacer. The performance improvement could be rationalized in terms of the increased absorption in the photoactive layer that arises from a better match of the spatial distribution of the light intensity to the maximum position of the polymeric P3HT:PCBM film.

3.3.3 Bulk Heterojunction

As mentioned in Sect. 3.3, due to the short exciton diffusion length in most organic materials (typically 3–40 nm), only photons absorbed within the exciton diffusion length from the donor–acceptor interface can survive the transport to the interface to subsequently dissociate into free charge carrier with a high yield. This invariably limits the thickness of the photoactive layer and thus the absorption efficiency. In addition, Yang and Forrest [53] recently reported that the η_{EQE} of devices with planar bilayer structure shows a strong dependence on exciton diffusion length. At larger photoactive layer thicknesses, absorption shifts toward the ITO–donor interface, where the narrow exciton distribution results in decreased η_{ED} and hence η_{EQE}. For standard CuPc/C_{60} bilayer device, the η_{ED} dropped from 0.6 in CuPc (10 nm)/C_{60} (20 nm) to 0.2 in CuPc (20 nm)/C_{60} (40 nm), and approached to 0 for total photoactive layer thickness increases to 100 nm.

The concept of bulk heterojunction circumvents the exciton diffusion limitation in the bilayer device structures. In 1995, Yu et al. [81] first utilize this approach by simply dispersing fullerene in a conjugated polymer and then spin coating the solution onto the device. Devices with the composite polymer film (C_{60}: MEH-PPV, where MEH-PPV is poly(2-methoxy-5-(2'-ethyl-hexyloxy)-1,4-phenylene vinylene)) showed a dramatically increased PCE of 2.9%, two orders of magnitude higher than that without the C_{60} dispersion. Bulk heterojunction consists of an intimate mixture of donor and acceptor materials via blending polymers or coevaporation methods. The mixed layer creates a spatially distributed donor–acceptor interface, where all excitons generated can be efficiently dissociated into free charge carriers throughout the bulk heterojunction, giving ~100% η_{ED}. Figure 17 depicts the schematic diagrams of the photovoltaic processes for both bilayer and bulk heterojunction devices. In addition, the bulk heterojunction essentially extends the photoactive layer thickness and decreases the charge carrier recombination probability, leading to a higher dark current and I_{SC}. Extensive studies had been carried out to investigate the charge carrier transport and photogeneration mechanisms within bulk heterojunction. Ultrafast spectroscopic analysis revealed that ultrafast charge transfer (~45 fs) from the conjugated polymers to fullerenes has been observed, resulting in a nearly unity charge transfer efficiency [82]. Yang and Forrest recently further supported that the exciton diffusion efficiency is equal to one for all photoactive layer thicknesses due to the highly distributed donor–acceptor interface [53]. This definitely provides a simple means to extend the photoactive layer thickness and thus increasing the absorption efficiency.

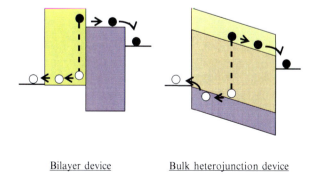

Bilayer device   Bulk heterojunction device

Fig. 17 Schematic diagram of photovoltaic processes for bilayer and bulk heterojunction devices

However, the morphology, or microstructure, of the mixed layer of bulk heterojunction plays a crucial role in determining the device efficiency. In 2001, Shaheen et al. [83] demonstrated that the photovoltaic responses of polymeric OPV devices were controlled by the morphology of the polymer layer. It was found that the conformation of polymer chains was controlled by the selection of the casting solvents, in which aromatic solvents (e.g., dichlorobenzene or chlorobenzene) could prevent the formation of isolated regions of polymer and PCBM in the film and enhance the polymer chain packing to increase hole mobility. Particularly, devices with chlorobenzene-casted film had a high I_{SC} of 5.25 mA cm^{-2}, a V_{OC} of 0.82 V, and a FF of 0.61, corresponding to a PCE of 2.5%. These values were comparatively higher than those of devices with toluene-casted film ($I_{SC} = 2.33$ mA cm^{-2}, $V_{OC} = 0.82$ V, FF $= 0.50$, and PCE $= 0.9\%$). Furthermore, the intrinsically low carrier mobilities in the bulk heterojunction are the bottleneck for maximizing the η_{EQE} of OPV devices. Rand et al. [84] found that in the CuPc:C_{60} mixture (1:1 by weight), the electron and hole mobilities were reduced by more than one order of magnitude compared to that of the neat CuPc or C_{60} films. In particular, hole mobility of the mixed film at zero electric field is 3.3×10^{-6} cm^2 V^{-1} s^{-1}, two orders of magnitude lower than that of the pristine CuPc film (7.4×10^{-4} cm^2 V^{-1} s^{-1}). This led to a significantly lower charge carrier collection efficiency ($\eta_{CC} \sim 0.4$), resulting from a large recombination rate at the increased interface area, as compared to that with planar device structure ($\eta_{CC} = 1.0$) [53].

3.3.4 Low Bandgap Photoactive Materials

One of the major obstacles for achieving high PCE is the limited spectral overlap between the absorption of photoactive materials and the solar spectrum, resulting in a low photocurrent. Indeed, ~62% of the total solar photon flux is at wavelengths $\lambda > 600$ nm with approximately 40% in the red and near-infrared (NIR) spectrum at $600 < \lambda < 1{,}000$ nm. However, the optical bandgap of most organic photoactive materials is not optimized with respect to the solar spectrum, in which only 20–30% of solar spectrum can be absorbed. For instance, CuPc, a commonly used donor material in OPV devices, has an absorption spectrum that falls off at $\lambda > 700$ nm [80, 84]. This suggests that new materials need to be developed that can absorb NIR radiation, and effectively convert the absorbed photons into electricity.

Various new organic materials with low optical bandgap and broad spectral coverage extending into the NIR radiation have been designed and synthesized. With the extensive device engineering efforts, encouraging power conversion efficiencies of 4–5% have been demonstrated for the NIR OPV devices [71, 73, 74, 85, 86]. Here, there are several examples of novel low optical bandgap organic materials that are promising candidates for further exploration in OPV devices with real applications. In 2006, Mühlbacker et al. [85] had designed and synthesized a new polymeric material, poly[2,6-(4,4-bis-(2-ethylhexyl)-4H-cyclopenta[2,1-b;3,4-b']-dithiophene)-alt-4,7-(2,1,3-benzothiadiazole)] (PCPDTBT). PCPDTBT

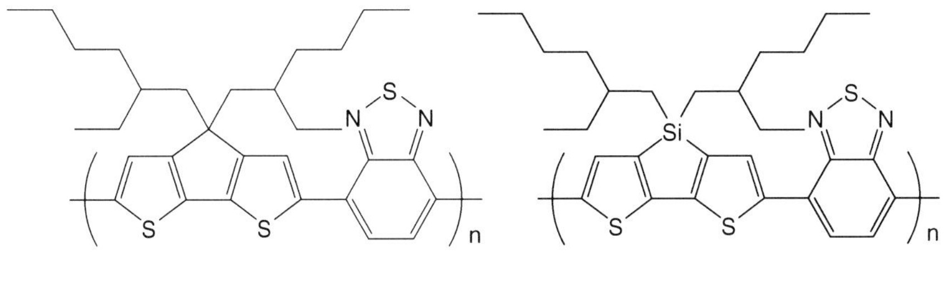

Fig. 18 Chemical structures of PCPDTBT and PSBTBT

was the first candidate of a new class of copolymers utilizing a cyclopentadithiophene unit as the donor block in the polymer chain, and it exhibited a low optical bandgap of 1.46 eV with spectral coverage extending up to 900 nm. In addition, the PCPDTBT/P_{70}BM-based polymeric OPV devices attained a high PCE of 3.2%. Later, Hou et al. [86] modified the PCPDTBT by substituting the 9-position of the fluorene units with silicon atoms in order to improve the hole-transporting property of the material and two 2-ethylhexyl side chains attached to the silicon atom to enhance the solubility to synthesize a novel silole-containing polymer, poly[(4,4'-bis(2-ethylhexyl)dithieno[3,2-b;2',3'-d]silole)-2,6-diyl-alt-2,1,3-benzothiadiazole-4,7-diyl] (PSBTBT). Figure 18 shows the chemical structures of PCPDTBT and PSBTBT. This polymer performed a high hole mobility of 3×10^{-3} cm^2 V^{-1} s^{-1}, that was three times higher than that for the unmodified polymer. In addition, such polymeric OPV device exhibited a high PCE of 5.1% under AM1.5G (G: Global) illumination with light intensity of 100 mW cm^{-2}.

Very recently, Heeger's group reported a polymeric OPV device with the world record PCE of 6.1% under 1 sun AM 1.5G irradiation [74]. This success relied on the use of alternating copolymer based on poly(2,7-carbazole) derivatives, with a suitable electron-deficient moieties to tune the electronic energy gap and absorption spectrum into the NIR region. In particular, the smaller bandgap of polymer can harvest a larger fraction of the solar radiation spectrum, while the deeper HOMO energy of the carbazole group can effectively increase the V_{OC} of the devices. Such polymeric OPV device attained a high I_{SC} of 10.6 mA cm^{-2}, high V_{OC} of 0.88 V, and FF of 0.66, resulting in an impressive PCE of 6.1%. More importantly, from the IPCE measurement, it was found that the internal quantum efficiency is close of 100%, implying that essentially every absorbed photon resulted in a separated pair of charge carriers and that all photogenerated carriers could be collected at the respective electrodes.

3.3.5 Tandem Structures

Since the first demonstration of tandem device structure by Kido et al. [87], this concept has been widely used in OLEDs for improving the device performance

[33–35]. Tandem structure consists of two or more individual subcells connected electrically in series via charge connecting unit. Under operation, charge carriers will be generated within the connecting unit and are injected into the adjoining electroluminescence (EL) elements. In principle, both luminous efficiency and operation lifetime can be proportionally increased with the number of EL units. This approach can also be employed to achieve white light by the combination of the two or more complementary colors.

This concept of tandem OPV structure had also been realized by Xue et al. [66], in which two individual OPV cells were connected in series by an ultrathin (~5 Å) discontinuous metallic film as an interlayer. The metal nanoclusters acted as efficient charge recombination centers that allowed these subcells to operate without incurring a voltage drop at the junction between two subcells. To obtain the series configuration, an inverse oriented heterojunction between the acceptor layer of one cell and the donor layer of the adjacent cell was employed. Such tandem OPV devices exhibited a high V_{OC}, close to the sum of the V_{OC} of two subcells, and high PCE of 5.7%. Kim et al. [88] also demonstrated a tandem polymeric OPV device by stacking two subcells (PCPDTBT:PCBM and P3HT:$PC_{70}BM$) together via a thin TiO_x interlayer. This resulted in an impressive PCE of 6.5%.

However, due to the optical interference effects and the thickness-dependent optical properties of the thin interlayer, realization of tandem structure needs careful optimization of the respective photoactive layer thickness of each subcell. In particular, the cell structure should be arranged for making the front cell rich in the longer wavelength absorbing materials and the back cell rich in the shorter wavelength materials [66, 88]. In addition, the current of the tandem OPV devices is determined by the current generation in the lowest subcell [88]. This significantly poses an upper limit on the PCE of tandem devices.

3.4 Challenges of OPV Devices

With the continuous efforts on the development of new organic materials and innovative device engineering, the PCE of OPV devices has been steadily improved up to 6.1%; however, this efficiency is still not sufficient to meet the realistic specifications for commercialization. The major obstacle for achieving high PCE is the mismatch of the absorption of photoactive materials to the terrestrial solar spectrum. In particular, the optical bandgap of most photoactive organic materials is much larger than the maximum photon flux at 1.8 eV, wasting over 60% of the solar spectrum. Scharber et al. [55] had systematically studied the dependence of V_{OC} on the bandgap and HOMO levels of 26 conjugated polymers and predicted the limited efficiency of bulk heterojunction OPV devices. It was found that PCE of 10% can theoretically be achieved for the PCBM-based polymeric OPV device, provided that the donor polymer must have a bandgap <1.74 eV and a LUMO level <-3.92 eV, assuming that the FF and the average η_{EQE} remain equal to 0.65. Meanwhile, Rand et al. [56] presented a comprehensive device model to predict the photovoltaic responses of small molecular-based OPV devices. This model infers that the maximum PCE as high as 12% is feasible for an optimized double heterostructure OPV device. However, its success relies on the development of new organic materials, in which the donor and acceptor materials should have optical bandgap of 1.5 and 1.8 eV, respectively, and a large interface offset energy, that is, $\Delta = I_P$ (donor) $- E_A$ (acceptor) $= 1.1$ eV. This suggests that new photoactive materials with low optical bandgap are definitely needed to be developed.

4 Conclusions

With the growing demand on energy consumption, there is an urgent need for implementing more energy saving measures. Many nations have committed on their new policies to implement the ways for saving energy and promoting it. Particularly, solid-state organic devices, especially WOLEDs and organic solar cells, are targeted as effective ways for reducing the energy consumption and developing renewable energy in the world. Because of high η_P and ultra-low costs of organic materials, WOLEDs and organic solar cells are gaining acceptance as the new generation of solid-state lighting sources to replace and complement the incandescent bulbs and an alternative type of solar cells to dislodge or complement the traditional Si solar cell for electricity generation, respectively. Indeed, WOLEDs with fluorescent tube efficiency as high as 102 lm W^{-1} and organic solar cells with high PCE of 6.1% have been achieved in the laboratory. With the continuous efforts on improving their lifetimes and reducing the manufacturing cost, WOLEDs and organic solar cells have a very bright future to be used widely.

Acknowledgments This work has been supported by the University Grants Committee Areas of Excellence Scheme (AoE/P-03/08) and the Strategic Research Theme on Molecular Materials of The University of Hong Kong.

References

1. US Energy Information Administration, International Energy Outlook 2009 (August 2009)
2. US Energy Information Administration, Annual Energy Outlook 2009 (March 2009)
3. D'Andrade BW, Forrest SR (2004) White organic light-emitting devices for solid-state lighting. Adv Mater 16:1585–1595
4. US Energy Information Administration, Annual Energy Review 2006 (June 2007)
5. US Department of Energy (2002) National lighting inventory and energy consumption estimate, vol 1. Navigant Consulting, Washington DC
6. Reinkek S, Lindner F, Schwartz G, Seidler N, Walzer K, Lüssem B, Leo K (2009) White organic light-emitting diodes with fluorescent tube efficiency. Nature 459:234–238
7. Su SJ, Gonmori E, Sasabe H et al (2008) Highly efficient organic blue- and white-light-emitting devices having a carrier- and exciton-confining structure for reduced efficiency roll-off. Adv Mater 20:4189–4194

8. Schwartz G, Reineke S, Walzer K et al (2008) Reducing efficiency roll-off in high-efficiency hybrid white organic light-emitting diodes. Appl Phys Lett 92:053311
9. Forrest SR, Bradley DDC, Thompson ME (2003) Measuring the efficiency of organic light-emitting devices. Adv Mater 15:1043–1048
10. Commission Internationale de L'éclairage (CIE) (1986) Colorimetry, Publication Report No. 15.2
11. Borbély Á, Sámson Á, Schanda J (2001) The concept of correlated color temperature revisited. Color Res Appl 26:450–457
12. Commission Internationale de L'éclairage (CIE) (1974) Method of measuring and specifying color rendering properties of light sources, Publication Report No. 13.2
13. Joint ISO/CIE Standard: CIE standard illuminants for colorimetry provide explanations and descriptions of the CIE standard illuminants. ISO 10526:1999/CIE S005/E-1998
14. D'Andrade BW, Holmes RJ, Forrest SR (2004) Efficient organic electrophosphorescent white-light-emitting device with a triple doped emissive layer. Adv Mater 16:624–628
15. Lee MT, Lin JS, Chu MT et al (2008) Improvement in carrier transport and recombination of white phosphorescent organic light-emitting devices using a composite blue emitter. Appl Phys Lett 93:133306
16. Eom SH, Zheng Y, Wrzesniewski E et al (2009) White phosphorescent organic light-emitting devices with dual triple-doped emissive layers. Appl Phys Lett 94:153303
17. Wang Q, Ding J, Ma D et al (2009) Harvesting excitons via two parallel channels for efficient white organic LEDs with nearly 100% internal quantum efficiency: fabrication and emission-mechanism analysis. Adv Funct Mater 19:84–95
18. Wang Q, Ding J, Ma D et al (2009) Highly efficient single-emitting-layer white organic light-emitting diodes with reduced efficiency roll-off. Appl Phys Lett 94:103503
19. Sun Y, Giebink NC, Kanno H et al (2006) Management of singlet and triplet excitons for efficient white organic light-emitting devices. Nature 440:908–912
20. Schwartz G, Fehse K, Pfeiffer M et al (2006) Highly efficient white organic light emitting diodes comprising an interlayer to separate fluorescent and phosphorescent regions. Appl Phys Lett 89:083509
21. Sun Y, Forrest SR (2007) High-efficiency white organic light emitting devices with three separate phosphorescent emission layers. Appl Phys Lett 91:263503
22. Schwartz G, Pfeiffer M, Reineke S et al (2007) Harvesting triplet excitons for fluorescent blue emitters in white organic light-emitting diodes. Adv Mater 19:3672–3676
23. Chang CH, Lin YH, Chen CC et al (2009) Efficient phosphorescent white organic light-emitting devices incorporating blue iridium complex and multifunctional orange-red osmium complex. Org Electron 10:1235–1240
24. Luo J, Li X, Hou Q et al (2007) High-efficiency white-light emission from a single copolymer: fluorescence blue, green, and red chromophores on a conjugated polymer backbone. Adv Mater 19:1113–1117
25. Liu J, Shao SY, Chen L et al (2007) White electroluminescence from a single polymer system: improved performance by means of enhanced efficiency and red-shifted luminescence of the blue-light-emitting species. Adv Mater 19:1859–1863
26. Thompson J, Blyth RIR, Mazzeo M et al (2001) White light emission from blends of blue-emitting organic molecules: a general route to the white organic light-emitting diode? Appl Phys Lett 79:560–562
27. D'Andrade BW, Brooks J, Adamovich V et al (2002) White light emission using triplet excimers in electrophosphorescent organic light-emitting devices. Adv Mater 14:1032–1036
28. Williams EL, Haavisto K, Li J et al (2007) Excimer-based white phosphorescent organic light emitting diodes with nearly 100% internal quantum efficiency. Adv Mater 19:197–202
29. Mazzeo M, Pisignano D, Della Sala F et al (2003) Organic single-layer white light-emitting diodes by exciplex emission from spin-coated blends of blue-emitting molecules. Appl Phys Lett 82:334–336

30. Palilis C, Mäkinen AJ, Uchida M et al (2003) Highly efficient molecular organic light-emitting diodes based on exciplex emission. Appl Phys Lett 82:2209–2211
31. Tong QX, Lai SL, Chan MY et al (2007) High-efficiency nondoped white organic light-emitting devices. Appl Phys Lett 91:023503
32. Kido J, Nakada T, Endo J et al (2002) High efficiency organic EL devices having charge generation layer. In: Neyts K, De Visschere P, Poelman D (eds) Proceedings of the 11th international workshop on inorganic and organic electroluminescence and 2002 international conference on the science and technology of emissive displays and lighting, Universiteit Ghent, Ghent, Belgium 2002, p 539
33. Liao LS, Klubek KP, Tang CW (2004) High-efficiency tandem organic light-emitting diodes. Appl Phys Lett 84:167–169
34. Tsutsui T, Terai M (2004) Electric field-assisted bipolar charge spouting in organic thin-film diodes. Appl Phys Lett 84:440–442
35. Lai SL, Chan MY, Fung MK et al (2007) Copper hexadecafluorophthalocyanine and copper phthalocyanine as a pure organic connecting unit in blue tandem organic light-emitting devices. J Appl Phys 101:014509
36. Chan MY, Lai SL, Lau KM et al (2007) Influences of connecting unit architecture on the performance of tandem organic light-emitting devices. Adv Funct Mater 17:2509–2514
37. Krummacher BC, Choong V-E, Mathai MK et al (2006) Highly efficient white organic light-emitting diode. Appl Phys Lett 88:113506
38. Ji W, Zhang L, Gao R et al (2008) Top-emitting white organic light-emitting devices with down-conversion phosphors: theory and experiment. Opt Express 16:15489–15494
39. NanoMarkets LC, OLED Lighting Markets 2008 (September 2008)
40. US Department of Energy, Solid-State Lighting Research and Development Portfolio: Technology Research and Development Plan FY'07-FY'12., Navigant Consulting, Inc. and Radcliffe Advisors (January 2007)
41. Optrex Europe GmbH homepage (http://www.optrex.de)
42. OSRAM homepage (http://www.osram-os.com/appsos/showroom/)
43. Global Research Blog website (http://ge.geglobalresearch.com/blog/worlds-first-demonstration-of-roll-to-roll-processed-oleds/)
44. eMercedesBenz website (http://www.emercedesbenz.com)
45. Yano Research Institute Ltd, World Solar Cell Market: Key Research Findings 2009 (http://www.yanoresearch.com). Figures are reproduced with permission from Yano Research Institute
46. LG Electronics website (http://www.lge.com)
47. Eindhoven University of Technology (2008, May 14). New world record for important class of solar cells
48. National Renewable Energy Laboratory (NREL) Newsroom, NREL of the US Department of Energy (http://www.nrel.gov/news/)
49. University of Delaware (2007, July 23). UD-led team sets solar cell record, joins Dupont on $100 million project
50. Wadell AL, Forrest SR (2006) High power organic solar cells from efficient utilization of near-infrared solar energy. Mater Eng News pp 10–11
51. Hoppe H, Sariciftci NS (2004) Organic solar cells: an overview. J Mater Res 19:1924–1945
52. Peumans P, Yakimov A, Forrest SR (2003) Small molecular weight organic thin-film photodetectors and solar cells. J Appl Phys 93:3693–3723
53. Yang F, Forrest SR (2008) Photocurrent generation in nanostructured organic solar cells. ACS Nano 2:1022–1032
54. Rand BP, Genoe J, Heremans P et al (2007) Solar cells utilizing small molecular weight organic semiconductors. Prog Photovolt Res Appl 15:659–676
55. Scharber MC, Mühlbacher D, Koppe M et al (2006) Design rules for donors in bulk-heterojunction solar cells – towards 10% energy-conversion efficiency. Adv Mater 18:789–794

56. Rand BP, Burk DP, Forrest SR (2007) Offset energies at organic semiconductor heterojunctions and their influence on the open-circuit voltage of thin-film solar cells. Phys Rev 75:115327
57. Nunzi JM (2002) Organic photovoltaic materials and devices. C R Phys 3:523–542
58. Rostalski J, Meissner D (2000) Monochromatic versus solar efficiencies of organic solar cells. Sol Energy Mater Sol Cells 61:87–95
59. Onsager L (1938) Initial recombination of ions. Phys Rev 54:554–557
60. Tang CW (1986) Two-layer organic photovoltaic cell. Appl Phys Lett 48:183–185
61. Peumans P, Forrest SR (2001) Very-high-efficiency double-heterostructure copper phthalocyanine/C_{60} photovoltaic cells. Appl Phys Lett 79:126–128
62. Uchida S, Xue J, Rand BP et al (2004) Organic small molecular solar cells with a homogeneously mixed copper phthalocyanine: C_{60} active layer. Appl Phys Lett 84:4218–4220
63. Vogel M, Doka S, Breyer Ch, Lux-Steiner MCh, Fostiropoulos K (2006) On the function of a bathocuproine buffer layer in organic photovoltaic cells. Appl Phys Lett 89:163501
64. Peumans P, Uchida S, Forrest SR (2003) Efficient bulk heterojunction photovoltaic cells using small-molecular-weight organic thin films. Nature 425:158–162
65. Xue J, Uchida S, Rand BP et al (2004) Asymmetric tandem organic photovoltaic cells with hybrid planar-mixed molecular heterojunctions. Appl Phys Lett 85:5757–5759
66. Shao Y, Sista S, Chu CW et al (2007) Enhancement of tetracene photovoltaic devices with heat treatment. Appl Phys Lett 90:103501
67. Mayer AC, Lloyd MT, Herman DJ (2004) Postfabrication annealing of pentacene-based photovoltaic cells. Appl Phys Lett 85:6272–6274
68. Yoo S, Domercq B, Kippelen B (2004) Efficient thin-film organic solar cells based on pentacene/C_{60} heterojunctions. Appl Phys Lett 85:5427–5429
69. Potscavage WJ, Yoo S, Domercq B et al (2007) Encapsulation of pentacene/C_{60} organic cells with Al_2O_3 deposited by atomic layer deposition. Appl Phys Lett 90:253511
70. Kim JY, Kim SH, Lee HH et al (2006) New architecture for high-efficiency polymer photovoltaic cells using solution-based titanium oxide as an optical spacer. Adv Mater 18:572–576
71. Liang Y, Wu Y, Feng D et al (2009) Development of new semiconducting polymers for high performance solar cells. J Am Chem Soc 131:56–57
72. Padinger F, Rittberger RS, Sariciftci NS (2003) Effects of postproduction treatment on plastic solar cells. Adv Funct Mater 13:85–88
73. Li G, Shrotriya V, Huang J et al (2005) High-efficiency solution processable polymer photovoltaic cells by self-organization of polymer blends. Nat Mater 4:864–868
74. Park SH, Roy A, Beaupré S et al (2009) Bulk heterojunction solar cells with internal quantum efficiency approaching 100%. Nat Photonics. doi:10.1038/NPHOTON.2009.69
75. Peet J, Kim JY, Coates NE et al (2007) Efficiency enhancement in low-bandgap polymer solar cells by processing with alkane dithiols. Nat Mater 6:497–500
76. Peumans P, Forrest SR (2004) Separation of geminate charge-pairs at donor-acceptor interfaces in disordered solid. Chem Phys Lett 398:27–31
77. Djurovich PI, Mayo EI, Forrest SR et al (2009) Measurement of the lowest unoccupied molecular orbital energies of molecular organic semiconductors. Org Electron 10:515–520
78. Hänsel H, Zettl H, Krausch G et al (2003) Optical and electrical contributions in double-heterojunction organic thin-film solar cells. Adv Mater 15:2056–2060
79. Stübinger T, Brütting W (2001) Exciton diffusion and optical interference in organic donor-acceptor photovoltaic cells. J Appl Phys 90:3632–3641
80. Chan MY, Lai SL, Lau KM et al (2006) Application of metal-doped organic layer both as exciton blocker and optical spacer for organic photovoltaic devices. Appl Phys Lett 89:163515
81. Yu G, Gao J, Hummelen JC et al (1995) Polymer photovoltaic cells – enhanced efficiencies via a network of internal donor-acceptor heterojunctions. Science 270:1789–1791

82. Brabec CJ, Zerza G, Cerullo G et al (2001) Tracing photoinduced electron transfer process in conjugated polymer/fullerene bulk heterojunctions in real time. Chem Phys Lett 340:232–236
83. Shaheen SE, Brabec CJ, Sariciftci NS et al (2001) 2.5% efficient organic plastic solar cells. Appl Phys Lett 78:841–843
84. Rand BP, Xue J, Uchida S et al (2005) Mixed donor-acceptor molecular heterojunctions for photovoltaic applications. I. Materials properties. J Appl Phys 98:124902
85. Mühlbacher D, Scharber M, Morana M et al (2006) High photovoltaic performance of a low-bandgap polymer. Adv Mater 18:2884–2889
86. Hou J, Chen HY, Zhang S et al (2008) Synthesis, characterization, and photovoltaic properties of a low bandgap polymer based on silole-containing polythiophenes and 2, 1, 3-benzothiadiazole. J Am Chem Soc 130:16144–16145
87. Kido J, Nakada T, Endo J et al (2002) High efficiency organic EL devices having charge generation layer. In: Neyts K, De Visschere P, Poelman D (eds) Proceedings of the 11th international workshop on inorganic and organic electroluminescence and 2002 international conference on the science and technology of emissive displays and lighting, Universiteit Ghent, Ghent, Belgium, p 539
88. Kim JY, Lee K, Coates NE et al (2007) Efficient tandem polymer solar cells fabricated by all-solution processing. Science 317:222–225

White-Emitting Polymers and Devices

Hongbin Wu, Lei Ying, Wei Yang, and Yong Cao

Contents

1 Introduction 38
2 Characterization of White Polymer Light-Emitting Devices 38
   2.1 Color Quality of White Polymer Light-Emitting Devices 38
   2.2 Characterization for WPLED Efficiency 40
3 White Light-Emitting Polymers 44
   3.1 WLEPs Containing Blue and Yellow–Orange Chromophores 45
   3.2 WLEPs Containing Blue-, Green-, and Red-Emitting Chromophores 50
4 White Polymer Light-Emitting Devices 57
   4.1 WPLEDs Using Dye-Dispersed Polymer Blend 57
   4.2 WPLEDs from All-Polymer Blends 64
   4.3 WPLEDs from Excimer/Exciplex 68
   4.4 WPLEDs from Multiple Emissive Layers 70
   4.5 WPLEDs with Functional Charge-Injection/Transport Layers 71
5 Conclusions 76
References 76

Abstract Interest in using white polymer light-emitting devices (WPLEDs) for flat-panel displays, back-lighting sources for liquid-crystal displays, and next-generation solid-state lighting sources is increasing in the past several years due to their rapid improvement in terms of efficiencies and color quality. In this chapter, the progress on the synthesis of white light-emitting polymers and the advancement of WPLEDs were summarized.

H. Wu, L. Ying, W. Yang, and Y. Cao (✉)
Key Laboratory of Specially Functional Materials, Institute of Polymer Optoelectronic Materials and Devices, South China University of Technology, Guangzhou 510640, People's Republic of China
e-mail: yongcao@scut.edu.cn

1 Introduction

White light-emitting devices based on organic small molecules or polymers (WOLEDs or WPLEDs) have drawn intense attention in both scientific and industrial communities due to their potential applications in full-color flat-panel displays, back-lighting sources for liquid-crystal displays and next-generation solid-state lighting sources [1, 2]. In practice, in order to achieve white emission, mixtures of the three red, green, and blue (RGB) primary colors or two complementary colors are typically required. Various approaches toward realizing WOLEDs or WPLEDs have been reported, including multilayer structures fabricated by consecutive evaporations of red, green, and blue light-emitting compounds that allow sequential energy transfer, multiple component emissive layers containing appropriate ratios of RGB phosphorescent or fluorescent dopants, polymer blends containing RGB emitting species, charge transfer exciplexes, or excimers to achieve broad emission and single-component layer utilizing a polymer with broad emission covering the white light spectrum.

For WOLEDs, their performance has been improved significantly in the last few years and as a matter of fact the PE of WOLEDs has already exceeded that of incandescent light bulb (12–17 lm W^{-1}) [3, 4]. Furthermore, the efficiencies of the state-of-the-art devices have been shown to approach that of fluorescent lamp (40–70 lm W^{-1}) [5, 6], offering organic semiconducting lighting a very bright and promising future. On the other hand, despite their lower efficiencies, WPLEDs are of particular interest due to their potential advantages over the vacuum-deposited counterparts, such as low cost of manufacturing using solution process, easy processability over large area size using spin-coating, ink-jet printing, and screen printing technology, the compatibility with flexible substrates, relative small amount of wasted materials, and better control in doping level.

The aim of this chapter is to present the general approaches to achieve white emission from polymer-based devices. The chapter is organized as follows: after this brief introduction, Sect. 2 provides a short description of colorimetric quantities (color quality) of WPLEDs and characterization for their performance. Advances of the synthesis of single white emission polymers and their devices will be presented in Sect. 3. In Sect. 4, advances of WPLEDs from host/guest system, all-polymer blend, exciplexes/excimers broad emission, multiple component emissive layers, and highly efficient WPLEDs with functional charge-injection/transport layers will be addressed. Finally, Sect. 5 is devoted to provide an outlook for WPLEDs for practical applications.

2 Characterization of White Polymer Light-Emitting Devices

2.1 Color Quality of White Polymer Light-Emitting Devices

One of the most important applications of WOLEDs or WPLEDs is solid-state lighting. Generally, the color quality of lighting source is assessed by two important colorimetric quantities: chromaticity or color temperature (CT) [7] and color rendering indices (CRI) [8]. The color temperature of a light source is determined by comparing its chromaticity with that of an ideal black-body radiator. The temperature [usually measured in Kelvin (K)] at which the heated black-body radiator matches the color of the light source is that source's color temperature. Counterintuitively, higher Kelvin temperatures (5,000 K or more) are "cool" (green–blue) colors, and lower color temperatures (2,700–3,000 K) "warm" (yellow–red) colors. Cool-colored light is considered better for visual tasks. Warm-colored light is preferred for living spaces because it is considered more flattering to skin tones and clothing. Color temperatures in the 2,700–3,600 K range are recommended for most general indoor and task lighting. As a matter of fact, people are usually comfortable doing task under either noon sunlight (CT about 5,000 K), cool white (typical CT of fluorescent and compact fluorescent lamps (CFL) are about 4,000 K), or warm white (CT of incandescent lamp is about 3,000 K) while candle light has CT about 1,800 K. Table 1 shows the color temperature for the common sources. CRI is a quantitative measurement (in 0–100 scale) of the ability of a light source to reproduce the colors of various objects faithfully in comparison with reference source (i.e., daylight), with 100 representing perfect color reproduction. In another word, CRI refers to color fidelity, the accurate representation of object colors compared to those same objects under a reference source. In order to provide high-quality white illumination, a CRI of 70 should be considered as minimum for good quality of white light, similar to sunlight. The CRI of sunlight is taken as reference and other light sources are evaluated in reference to this CRI value. The quality of light source in terms of CRI values are 75–100 (excellent), 60–75 (good), 50–60 (fair), and 0–50 (poor and not suitable for critical application). Incandescent bulb has a CRI of 95 and is regarded as best artificial light. Table 2 depicts the CRI and correlated color temperature (CCT) for common white light sources.

The color of a light emitter can be defined by a CIE (x, y) coordinate, where CIE stands for Commission internationale de l'Eclairage (French name) or International Commission on Illumination (English name) [9], while the color appearance of any passive (reflective, non-emissive) material or any transmissive filter (e.g., stained glass) can be defined by a CIE (x, y) coordinate only under a given source of illumination. Thus, any color can be readily expressed in the CIE 1931 chromaticity diagram (Fig. 1). The inverted-U-shaped locus boundary represents monochromatic

Table 1 Color temperature for common sources [7]

Source	Color temperature (K)
Match flame	1,700
Candle flame	1,850
Incandescent light bulb	2,800–3,300
Studio "CP" light	3,350
Studio lamps, photofloods, and so on	3,400
Moonlight, xenon arc lamp	4,100
Horizon daylight	5,000
Typical daylight, electronic flash	5,500–6,000
Daylight, overcast	6,500
CRT screen	9,300

Table 2 CRI and CCT for common white light sources [8]

Light source	CRI	CCT (K)
Low-pressure sodium	~5	1,800
Clear mercury vapor	17	6,410
High-pressure sodium	24	2,100
Coated mercury vapor	49	3,600
Halophosphate warm white fluorescent	51	2,940
Halophosphate cool white fluorescent	64	4,230
Tri-phosphor warm white fluorescent	73	2,940
Halophosphate cool daylight fluorescent	76	6,430
Quartz metal halide	85	4,200
Tri-phosphor cool white fluorescent	89	4,080
Ceramic metal halide	96	5,400
Incandescent/halogen light bulb	100	3,200

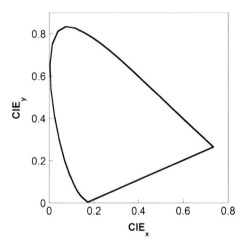

Fig. 1 CIE 1931 chromaticity diagram

visible light (450–650 nm), or spectral colors (loosely rainbow colors), while the lower bound of the locus is known as the line of purples and represents non-spectral colors obtained by mixing light of red and blue wavelengths. Colors on the periphery of the locus are saturated; colors become progressively desaturated and tend toward white somewhere in the middle of the plot. The point at $x = y = z = 0.333$ represents the white perceived from an equal-energy flat spectrum of radiation. Any color within a triangle defined by three primaries can be created (or recreated) by additive mixing of varying proportions of those primary colors. Typical values of CIE (x, y) coordinates for a series of white sources (defined as CIE Illuminants by CIE for calculation of color measurement) are summarized in Table 3 [10].

2.2 Characterization for WPLED Efficiency

The efficiency of WPLEDs is characterized by its luminous efficiency (LE), quantum efficiency (QE), and power efficiency (PE). Of which, LE and QE are

Table 3 Typical values of CIE (x, y) coordinates and color correlated temperature for CIE illuminants [10]

CIE illuminates	Note	CIE (x)	CIE (y)	CCT (K)
A	Incandescent/tungsten	0.448	0.407	2,856
D50	Horizon light	0.346	0.359	5,003
D55	Mid-morning/mid-afternoon daylight	0.332	0.347	5,503
D65	Noon daylight: television, sRGB color space	0.313	0.329	6,504
D75	North sky daylight	0.299	0.315	7,504
9,300 K	CRT monitor	0.285	0.293	9,300
E	Equal energy	0.333	0.333	5,454
F2	Cool white fluorescent	0.372	0.375	4,230
F5	Daylight fluorescent	0.314	0.345	6,350
F7	D65 simulator, daylight simulator	0.313	0.329	6,500

important for material evaluation while PE is important for device evaluation and engineering design. LE is measured in candela per ampere (cd A^{-1}) and obtained on the basis of measurement of luminous intensity (in candela, cd), or luminance (L, in candela per meter square, cd m^{-2}) at a given current density (J) by equation LE = L/J. It is important to note at a given radiation power, eye response to the incident light is strongly depended on the wavelength within the visible spectrum. As a result of eye response, luminance is usually defined differently from electromagnetic radiation at all other wavelengths. The response of the eye is represented by the luminosity function (Fig. 2). In other words, luminous intensity is the power emitted by a light source in a particular direction, weighted by the luminosity function. As shown in Fig. 2, monochromatic light at a wavelength of 555 nm (green) with an incident power of 1 W produces the highest response of 683 lm, while in contrast, monochromatic light at a wavelength of 630 nm (red) and 460 nm (blue) with the same power can only produce 181 lm and 61 lm, respectively. Usually, luminance is measured by a calibrated photodiode or a photometer.

OLEDs or PLEDs are a kind of charge carrier injection device which can convert electricity into light. So, naturally, the QE of a device is defined as the ratio between generated photons and the injected electron–hole pairs in the device. Of which, photons emitted outside the device are correlated to external quantum efficiency (EQE) while all the photons generated in the device contribute to its internal quantum efficiency (IQE). Accurate measurement of the EQE of a device can be achieved by measuring the total light output in all directions in an integrating sphere, while IQE can be deduced from EQE if extraction or out-coupling factor is determined. According to spin statistics, a ratio between singlet and triplet of 1:3 is formed when OLEDs are electrically excited, although the ratio of singlet-to-triplet is still a topic of intense debates [11, 12]. As a result, the theoretical maximum IQE and EQE of OLEDs based on fluorescent organic materials which use only singlet excitons are 25% and 5%, respectively, if an out-coupling factor of 25–30% is encountered [13]. This leads to an upper limit of IQE of 25% for fluorescent emitters. In contrast, OLEDs based on phosphorescent dopants can

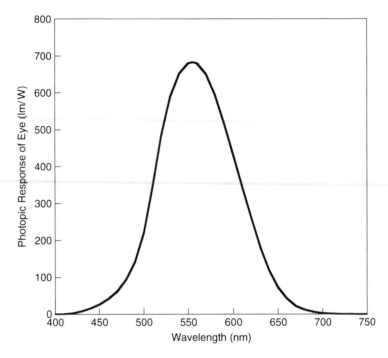

Fig. 2 The response of a typical human eye to light (luminosity function)

allow for a conversion of up to 100% of injected charges; both singlet and triplet excitons can be harvested into emitted photons, leading to a theoretical IQE of 100% [14, 15], much higher than those of their fluorescent counterparts.

Artificial white light sources are usually evaluated in terms of PE, which is defined as the luminous flux output (in lumen) per input power of the device, measured in lumen per watt (lm W^{-1}). For light sources like OLEDs or PLEDs emitting into a half plane, the relationship between luminous intensity (in candela) and luminous flux (in lumen) follow 1 cd = π lm [16, 17], so one can readily obtain PE by equation PE = $\pi \times$ LE/V, where V is the applied voltage. It is clearly shown in Fig. 2 that the upper limit of PE for monochromatic emission is 683 lm W^{-1} (555 nm green), while theoretically the maximum efficiency of a white light source is about 180–200 lm W^{-1} [18]. It is also important to point out that, when WOLEDs or WPLEDs are used as light source, in general, all of the emitting photons should be taken into account for illumination if they were redirected to the forward-viewing direction by some lighting fixtures. That is, in order to get the total LE and PE, a factor of 1.7–2.3 should be applied to the forward-viewing LE and PE measured by a photodiode or photometer [19]. Typical values of PE for various light sources are listed in Table 4 for comparison [20].

As can be seen from Table 4, for traditional light sources such as incandescent bulbs, their PE are typically between 12 and 17 lm W^{-1} as 90% of input power is wasted and is converted into heat. So far, the most widely used domestic lighting

Table 4 Typical/maximal value of power efficiency for various light sources [20]

Category	Type	Power efficiency (lm W^{-1})
Combustion	Candle	0.3
	Gas mantle	2
Incandescent	40 W tungsten incandescent (120 V)	12.6
	100 W tungsten incandescent (220 V)	13.8
	100 W tungsten incandescent (120 V)	16.8
	Quartz halogen (12–24 V)	24
	Photographic and projection lamps	35
Fluorescent	5–24 W compact fluorescent	45–60
	T12 tube with magnetic ballast	60
	T5 tube	70–100
	T8 tube with electronic ballast	80–100
Light-emitting diode	White LEDs (inorganic)	10–160
	White LEDs (organic)	1–102
Arc lamp	Xenon arc lamp	30–50
	Mercury-xenon arc lamp	50–55
Gas discharge	1,400 W sulfur lamp	100
	Metal halide lamp	65–115
	High-pressure sodium lamp	150
	Low-pressure sodium lamp	183–200
Theoretical maximum for monochromatic	555 nm monochromatic light	683
Theoretical maximum for white emission		180–300

source is fluorescent lamp, which is characterized by high PE (45–100 lm W^{-1}), four to six times more efficient than incandescent bulbs, and with excellent color and low price. Furthermore, fluorescent lamp can be used for both main electricity (110/220 V AC) and 12 V DC operation. Especially, in the near several decades, CFL were invented to replace incandescent lamp and can fit in the existing light screw fixtures formerly used for incandescent lamp. Despite these advantages, fluorescent lamp contains small but significant amounts of toxic mercury in the tube, which may complicate the disposal of fluorescent lamps.

As a rapid developing technology, WLEDs based on inorganic semiconductor for solid-state lighting are also very attractive because they can provide very high efficiency as a result of their capabilities to convert more than 90% of input power into photons. Recently, the PE of inorganic white LEDs starts to surpass that of CFL, and their color quality is improving too. As a result of this, inorganic white LEDs have already been found in the market and have been used in many lighting applications, such as traffic signal lights and portable light sources. However, the major challenges facing this technology for solid-state lighting are their high cost and the relatively small power output (usually 1–5 W and 100–300 lm).

In contrast to small (point-like) light sources typically from inorganic GaN crystals, WOLEDs or WPLEDs can be applied as large-area light sources due to inexpensive processing of amorphous thin organic or polymer films on large-area substrate. In recent years, PE of WOLEDs based on thermal deposition technology significantly improved to a level that is comparable to their inorganic counterparts based on GaN LEDs due to advances in novel materials synthesis and optimized device structure in the past few years. Especially, the recent development in using phosphorescent materials had led to significant improvement in device efficiency. For example, D'Andrade et al. [19] demonstrated that by simultaneously doping a wide gap host p-bis(triphenylsilyl)benzene (UGH2) with RGB-emitting iridium complexes (iridium(III) bis(2-phenylquinolyl-$N,C^{2'}$) acetylacetonate [PQIr] for red, fac-tris(2-phenylpyridine)iridium [Ir(ppy)$_3$] for green, and bis(4′,6′-difluorophenylpyridinato)-tetra-kis(1-pyrazolyl)borate iridium(III) [FIr6] for blue, respectively), the device had a high peak total PE of (42 ± 4) lm W^{-1} (corresponding to a PE of (26 ± 3) lm W^{-1} for forward viewing) at ~3 × 10^{-3} mA cm^{-2}, falling to (14 ± 2) lm W^{-1} at a higher current density of 10 mA cm^{-2}, with CIE coordinates of (0.40, 0.46) at 1.0 mA cm^{-2} and (0.38, 0.45) at 10 mA cm^{-2}, respectively. In addition, Kido et al. reported a highly efficient WOLED with a forward-viewing PE of 53 lm W^{-1} (can be as high as 70–100 lm W^{-1} if photons emitted in all directions can be redirected) at a typical luminance of 100 cd m^{-2} via a carrier- and exciton-confining structure [5]. So far, the best PE of WOLEDs based on small organic molecules is about 102 lm W^{-1} (total PE based on photons emitted in all directions), 50% of the theoretical maximum for white source [6], representing comparable efficiency with the best inorganic white LEDs.

Although the PE of most of WPLEDs based on solution processing is substantially lower than that from GaN LEDs, the PE of WPLEDs had recently been demonstrated to exceed that of incandescent light bulbs (12–17 lm W^{-1}) [21–24], representing a significant advance toward practical applications.

3 White Light-Emitting Polymers

Polymer light-emitting devices that can emit white light are particularly useful in making backlights for liquid-crystal displays or flat-panel displays. However, conventional PLEDs can only produce electroluminescence with the full width at half maximum (FWHM) of within a range of about 100 nm. In general, the approach to realize white emission covering the entire visible range from 400 to 700 nm is to use polymer/polymer blend systems, or it can be carried out by doping a large-bandgap polymer host with various narrow-bandgap dyes (fluorescent and/or phosphorescent) (see Sect. 4). In order to overcome possible problem of phase segregation at high current densities or long shelf lifetime caused by dopant/host blend system, synthetic chemists took strategy to design and synthesize novel single white light-emitting polymers (WLEPs) with various red- and green-, yellow-, or orange-emitting components chemically attaching into large-bandgap polymer host molecules. By decreasing the content of the incorporated chromophores to a certain level, unlike complete energy transfer for monochromatic emission, incomplete energy transfer will take place, resulting in a simultaneous emission from the host and the chormophores. Thus, the spectral range of each individual color overlaps with each other, creating a continuous spectrum very close to standard white light. Single WLEPs principally can be obtained by incorporating two (blue and orange) chromophores or three (blue, green, and red) chromophores into a polymer chain. Obviously, the latter is more difficult to synthesize but potentially have better color quality while the former type of white light-emitting polymer is easier to be synthesized.

3.1 WLEPs Containing Blue and Yellow–Orange Chromophores

3.1.1 Complementary Colors Based WLEPs Utilizing Singlet Exciton

In principle, white electroluminescent polymers containing complementary blue and orange chromophores have great advantages in term of their ease of synthesis and potentially higher EL efficiencies due to stronger response in the luminosity function (Fig. 2). By chemically incorporating an orange-emissive dopant (1,8-naphthalimide) unit into the main chain of a blue-emitting polymer host polyfluorene, a novel single-component polymer system (**P1**) with two individual emission species and simultaneous blue and orange emission to give white light emission was developed [25]. An LE of 5.3 cd A^{-1} and a PE of 2.8 lm W^{-1} at 6 V were obtained. Although the CIE coordinates (0.25, 0.35) of the device deviated from standard white light emission, consistent with a molecular dispersion of chromophore molecules in a polymer matrix, the WPLEDs from the copolymer showed a very stable white light emission at different driving voltages and brightnesses. Therefore, this strategy seems to be a general route for developing WPLEDs with high efficiency, high brightness, and stable emission color from a single-component polymer.

Our group has reported a single-component white emission polymer consisting of a narrow-bandgap chromophore, orange-emitting 2,3-dimethyl-5,8-dithien-2-ylquinoxaline (DDQ) unit in the backbone (**P2**) [26]. By controlling the feed ratio precisely, a small amount of DDQ units was incorporated into the polymer backbone, leading to balanced blue emission from the backbone of PFO and orange–red emission from the DDQ units, respectively. A maximum LE of 4.1 cd A^{-1} and CIE coordinates of (0.31, 0.28) had been achieved. A particular advantage of the obtained PFO-DDQ polymer is that color coordinates are extremely stable and remain constant in a wide range of operating current densities, which is a very important and desirable feature for display and lighting applications. Similarly, incorporation of an orange-emitting derivative of a widely used arylidene laser dye 4-(dicyanomethylene)-2-methyl-6-[*p*-(dimethylamino)styryl]-4*H*-pyran (DCM) into the main chain of polyfluorene can also lead to white electroluminescence (**P3**) [27]. Devices based on copolymer with optimized composition of orange chromophore exhibit almost balanced blue emission peaked at 423 and 450 nm and orange emission at 580 nm, and a nearly pure white CIE coordinates of (0.33, 0.31) was achieved.

Another copolyfluorene derivative was synthesized by the incorporation of 0.1 and 0.025 mol% orange 2,5-dihexyloxy-1,4-bis(2-thienyl-2-cyanovinyl) benzene chromophore for white electroluminescence through incomplete energy transfer (**P4**) [28]. The copolymer films showed PL peaks at about 428 and 570 nm originating from fluorene segments and the orange chromophore, respectively. The EL devices showed a broad emission band, covering the entire visible region, with CIE coordinates of (0.36, 0.35) and (0.32, 0.30) for 0.1 and 0.25 mol% of devices, respectively. The device based on the copolymer containing 0.25 mol% of orange chromophore exhibited a maximal LE of 1.98 cd A^{-1} and a peak luminance of 3,011 cd m^{-2}.

Also, Liu et al. introduced another kind of orange unit of quinacridone, which has very high PL QE of unity, into the main chain of polyfluorene (**P5**) [29]. With the increase of the quinacridone unit's content in the copolymers, it was found that the relative intensity of the orange emission band in the EL spectra becomes stronger owing to the higher degree of complete energy transfer and charge trapping. Thus, white emission can be readily realized by adjusting the content of the quinacridone unit, tuning the relative intensity of the blue emission from polyfluorene and the yellow emission from quinacridone. Particularly, WPLEDs emitted white light with CIE coordinates of (0.27, 0.35), with an LE of 3.47 cd A^{-1} and a PE of 2.18 lm W^{-1} could be obtained.

Chien et al. reported white-emitting polymer in which orange-emitting benzoselenadiazole units were covalently incorporated into the backbone of a PF copolymer containing hole-transporting triphenylamine and electron-transporting oxadiazole pendant groups (**P6**) [30]. By carefully controlling the concentrations of the low energy emitting species in the resulting copolymers, a partial energy transfer from the blue fluorescent PF backbone to the orange fluorescent segments resulted in two balanced blue and orange emissions simultaneously. White electroluminescence accompanied by a maximal LE of 4.1 cd A^{-1} with CIE coordinates of (0.30, 0.36) was achieved.

So far, the most commonly used narrow-bandgap comonomers for WLEPs were aromatic derivatives containing heterocycles such as thiophene, bithiophene, benzothiazole, and benzodithiazole. Recently, Park et al. synthesized a new white-emitting copolymer by incorporating a phenothiazine derivative, 4,7-bis(6-bromo-3-thiophenyl-10-n-butylphenothiazine)-2,1,3-benzothiadiazole (BPTR) into the main chain of poly[9,9-bis(4-octyloxyphenyl)-fluorene-2,7-diyl] (**P7**) [31]. The near-white electroluminescence with LE of 1.84 cd A^{-1} and CIE coordinates of (0.27, 0.40) was afforded by polymers with low content (0.1 mol%) of BPTR moiety.

To obtain an efficient, ideal white emission from single-component polymer, it is important to have both balanced emission intensity and good matching of emission wavelengths. From this perspective, Tu et al. developed a series of white-emitting copolymers by incorporating 1,8-naphthalimide derivatives with different emission wavelengths into polyfluorene main chain (**P8, P9, P10**) [32]. It was found that the emission is very sensitive to the peak wavelength of the 1,8-naphthalimide derivatives. For example, at the same concentration of 0.5 mol% of the yellow component, the peak wavelength varied from 545 to 555 nm and 567 nm, and resulted in the emission color shift from (0.34, 0.49) to (0.29, 0.33) and (0.32, 0.36). The device based on **P10** showed the best performance with LE of 3.8 cd A^{-1} and PE of 2.0 lm W^{-1}.

Aiming at a new single-component polymer system with improved EL efficiencies for displays and lighting applications, Wang's group incorporated a highly efficient orange chromophore 4,7-bis(4-(N-phenyl-N-(4-methylphenyl)-amino) phenyl)-2,1-3-benzothiadiazole (TPABT) into the main chain of blue-emitting PF (**P11**) [33]. The single-layer PLEDs based on polyfluorenes (PL QE, $\Phi_{PL} = 0.55$) containing efficient TPABT unit ($\Phi_{PL} = 0.76$) gave a pure white electroluminescence with CIE coordinates of (0.35, 0.32) and a maximal LE of 8.99 cd A^{-1}, which was approximately two times more efficient than that utilizing a single polyfluorene containing 1,8-naphthalimide moieties ($\Phi_{PL} = 0.25$). However, for the copolymer (**P12**) based on the same orange chromophore of TPABT and less-efficient blue-emitting host poly(9,9-dioctyl-2,7-fluorene-alt-2,5-bis(hexyloxy)-1,4-phenylene) (PFB, $\Phi_{PL} = 0.30$), the device gave a much lower LE of 2.53 cd A^{-1}. These results indicated that both of the efficient blue- and orange-emissive species are necessary to achieve high EL efficiencies.

Apart from covalently incorporating orange-emitting species into the backbone, WLEPs can be obtained by grafting a small amount of orange-emitting chromophore onto the side chain of the blue-emitting polymer host with an alkyl spacer. Based on this strategy, Wang's group had developed two highly efficient WLEPs by covalently attaching an orange-emitting unit 4-(4-alkyloxy-phenyl)-7-(4-diphenylamino-phenyl)-2,1,3-benzothiadiazole (MOB-BT-TPA) (**P13**) or 4-(4-alkyloxy-phenyl)-7-(5-(4-diphenylamino-phenyl) -thiophene-2-yl)-2,1,3-benzothiadiazole (MOB-BT-ThTPA) (**P14**) unit to the side chain of polyfluorene with an alkyl spacer, respectively [34]. Compared with incorporating the orange chromophore into the main chain of polyfluorene, covalent attachment to the side chain (**P13**) can avoid the influence on the electronic properties of the host polymers. By adjusting the content of the orange chromophore, charge trapping on the orange unit and incomplete energy transfer from polyfluorene to the fluorescent MOB-BT-TPA can take place, leading to a simultaneous emission from the orange chromophore ($\lambda_{max} = 545$ nm) and the host polyfluorene ($\lambda_{max} = 432$ and 460 nm), respectively, and thus resulting in a white electroluminescence with an LE of 10.66 cd A^{-1} and CIE coordinates of (0.30, 0.40). In contrast, polymer (**P14**), in which MOB-BT-TPA was incorporated in the polyfluorene main chain, exhibited a lower efficiency of 7.30 cd A^{-1} and CIE coordinates of (0.29, 0.37). In order to further improve LE, Wang's group incorporated a highly efficient sky-blue-emitting chromophore dimethylamino naphthalimide (DMAN, $\Phi_{PL} = 0.84$) into the polyfluorene side chain (**P16**) [35]. In particular, single-layer device based on the DMAN-containing polyfluorene (**P16**) showed an LE of 12.8 cd A^{-1}, a PE of 8.5 lm W^{-1}, and CIE coordinates of (0.31, 0.36) (i.e., control devices based on polyfluorene (**P15**) only exhibited an LE of 9.3 cd A^{-1}).

From a synthetic chemistry perspective, it is fairly interesting to develop WLEPs with novel architecture instead of the widely employed linear structure. With respect to the topology and molecular design, WLEPs with two emissive species can be linear, end-capped as well as star-shaped or branched structures. Recently, star-like WLEPs with an orange unit 4,7-(4-(diphenylamino)phenyl)-2,1,3-benzothiadiazole (TPABT) as core and four blue-emitting polyfluorene arms were reported (**P17**) [36]. One of the advantages of these star-shaped WLEPs lies in the fact that the intermolecular interaction can be effectively suppressed due to the highly branched and globular molecular features. By adjusting the content of the orange-emitting core, white EL with simultaneous blue emission (λ_{max} = 420 and 440 nm) and orange emission (λ_{max} = 562 nm) was obtained with an LE of 7.06 cd A^{-1}, a PE of 4.4 lm W^{-1}, and CIE coordinates of (0.35, 0.39). Scheme 1 shows the chemical structures of complementary colors based WLEPs utilizing singlet excitons.

3.1.2 Complementary Colors Based WLEPs Utilizing both Singlet and Triplet Excitons

In all aforementioned approaches, only singlet excitons are utilized during the electroluminescent processes. In considering that phosphorescent materials can realize nearly 100% IQE by harvesting both singlet and triplet excitons, the generation of white emission by employing phosphorescent materials is of special interest. Among the reported phosphorescent complexes, iridium-based complexes have attracted particular interest because of their extremely high phosphorescent quantum efficiencies.

In order to incorporate triplet emitter into WLEPs, our group reported the realization of efficient white emission from a copolymer in which orange-red-emitting iridium complex, iridium(III) bis(2-(2-phenyl-quinoline-$N,C^{2'}$)-(2,4-acetylacetonate) (Phq), was covalently attached into the side chain of polyfluorene by a long alkyl spacer (**P18**) [37]. As a result of incomplete quenching of the blue emission from the host polyfluorene and the orange-red emission from the pendant iridium complex, white electroluminescence with an LE of 3.2 cd A^{-1} and CIE coordinates of (0.33, 0.32) was obtained.

Similarly, efficient WLEPs can also be achieved by grafting other orange-red-emitting iridium complexes to polyfluorenes. For instance, by incorporating two orange phosphorescent iridium complexes, iridium(III) bis(4-phenylquinoline-$N, C^{2'}$) (2,4-pentadionate) [PIr] and iridium(III) bis(4-methylquinoline-$N,C^{2'}$) (2,4-pentadionate) [MIr], into the side chain of polyfluorene (**P19, P20**), respectively, Mei et al. reported white electroluminescence with an LE of 4.49 cd A^{-1} and a PE of 2.35 lm W^{-1} with CIE coordinates of (0.46, 0.33) (for [PF–PIr]-based devices) and an LE of 0.83 cd A^{-1} and PE of 0.37 lm W^{-1} with CIE coordinates of (0.36, 0.32) (for [PF–MIr]-based devices) [38]. The CIE coordinates were improved to (0.34, 0.33) when copolymer [MIr10PF] (1.0 mol% feed ratio of [Br–MIr]) was employed as the white-emissive layer.

Furuta et al. demonstrated that a copolymer (**P21**), containing electron-transporting, hole-transporting, and phosphorescent platinum(II) complex moieties,

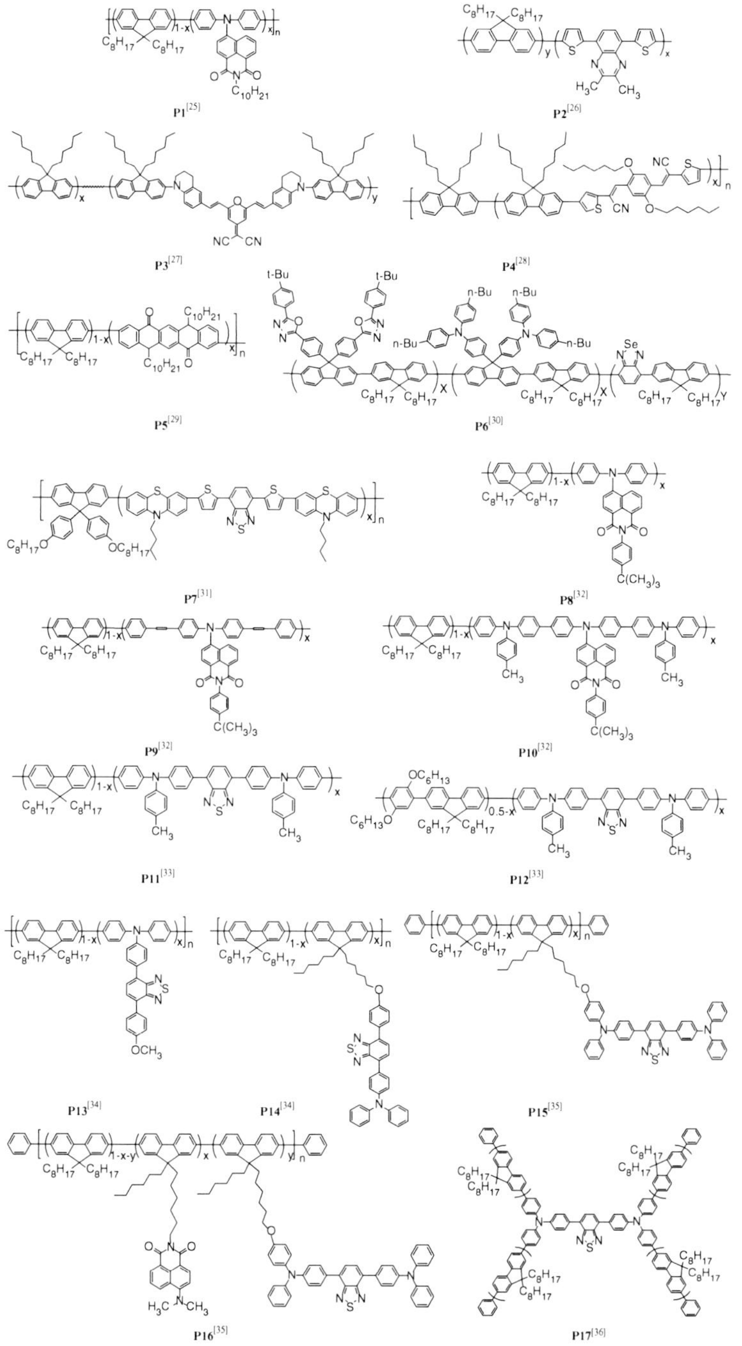

Scheme 1 Chemical structures of complementary colors based WLEPs utilizing singlet excitons

could be spin cast from solution to fabricate near-white emitting devices, in which white light was achieved from a combination of monomer emission and excimer/aggregate emission [39, 40]. Devices based on this copolymer showed an EQE up to 4.6% with CIE coordinates of (0.33, 0.50). It is interesting to note that the ratio of monomer/aggregate contributions in the EL spectra was invariant with the applied voltage, leading to a voltage-independent, good quality, and near-white broad emission. Scheme 2 shows the chemical structures of complementary colors based WLEPs utilizing both singlet and triplet excitons.

3.2 WLEPs Containing Blue-, Green-, and Red-Emitting Chromophores

Compared with white electroluminescent polymers based on blue–orange two complementary colors, the synthesis of white electroluminescent polymers with

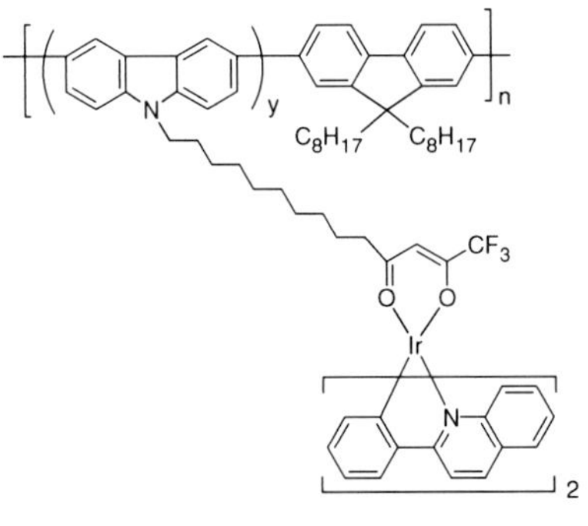

P18[37]

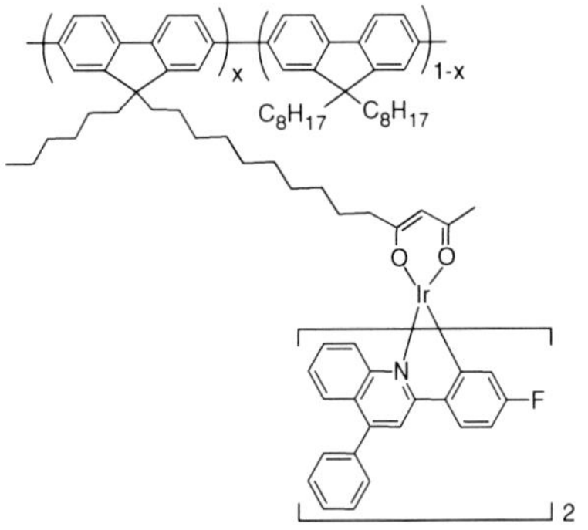

P19[38]

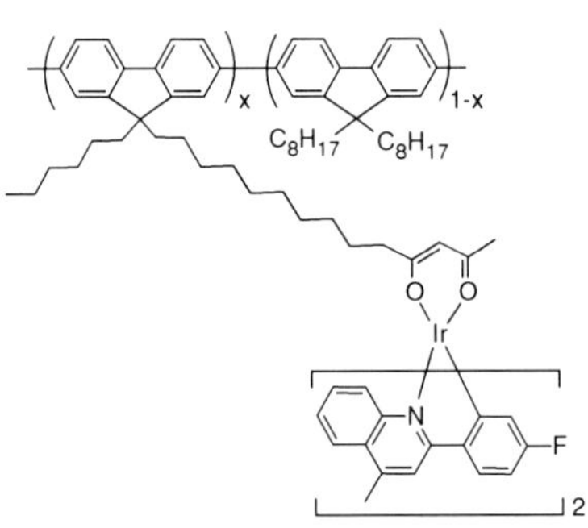

P20[38]

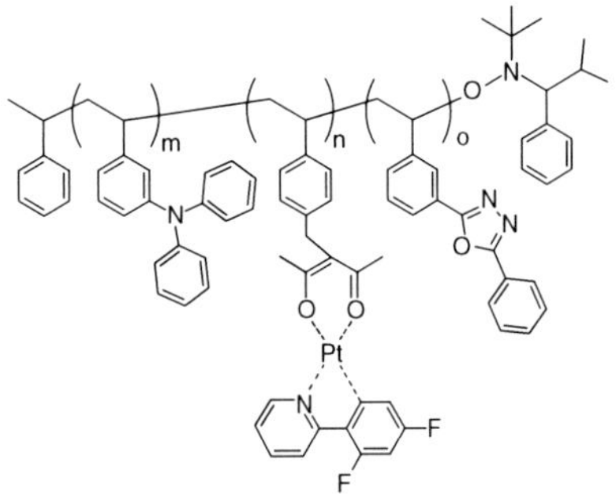

P21[39]

Scheme 2 Chemical structures of complementary colors based WLEPs utilizing both singlet and triplet excitons simultaneous RGB emission is more difficult. However, from a practical application perspective, for example, in order to meet the requirement of applications for full-color displays and provide good color quality for solid-state lighting, it is more desirable that WPLEDs from a single-component polymer can display simultaneous RGB emission.

3.2.1 RGB Primary Colors Based WLEPs Utilizing Singlet Excitons

Lee et al. reported the synthesis of a series of fluorene-based copolymer containing a red-emitting unit, 2-{2-(2-(2-[4-{bis(4-bromophenyl)amino}phenyl]-vinyl)-6-*tert*-butylpyran-4-ylidene}malonitrile (TPDCM), a green-emitting unit, {4-(2-[2,5-dibromo-4-{2-(4-phenylaminophenyl)vinyl}phenyl]-vinyl)phenyl}-diphenylamine (DTPA), and a blue-emitting unit, 2,7-dibromo-9,9-dihexylfluorene (DHF) (**P22**) by Ni-mediated polymerization [41]. By adjusting the concentration of the red and green chromophore, the emission color can be readily tuned. As a result, the optimized white electroluminescent copolymer was obtained when 3 mol% green-emitting DTPA and 2 mol% red-emitting TPDCM were incorporated. Device with an LE of 0.1 cd A^{-1} and a peak luminance of 820 cd m^{-2} was realized with CIE coordinates of (0.33, 0.35).

Soon after that, Wang and coworkers adopted a slightly different synthetic strategy by which a green-emitting component was attached to the pendant chain and a red-emitting component was incorporated into the blue-emitting polyfluorene backbone (**P23**) [42]. By carefully adjusting the content of the red- and green-emissive components, as a result of incomplete energy transfer from the blue macromolecule to the red and green components, simultaneous blue (445 nm), green (515 nm), and red (624 nm) emission from a single-component polymer could be achieved. Such EL device exhibited an LE of 1.6 cd A^{-1}, a PE of 0.8 lm W^{-1}, and CIE coordinates of (0.31, 0.34) at bias of 8–12 V.

Our group also proposed a similar strategy to realize white electroluminescence simultaneously with three primary color emissions from a single conjugated copolymer with a backbone consisting of PFO, 2,1,3-benzothiadiazole (BT), and 4,7-bis(2-thienyl)-2,1,3-benzothiadiazole (DBT) as the blue-, green-, and red-emitting units, respectively (**P24**) [43]. Incomplete energy transfer from the blue polyfluorene backbone to the green- and red-emitting units, in combination with restricted energy transfer from the green to the red due to effective isolation of these units along the conjugated backbones, resulted in three distinguishable primary emissions, thereby giving a broad white emission of balanced RGB components with desired CIE coordinates. The device fabricated from the copolymer PFO-R010-G018, which contained 0.01 mol% DBT and 0.018% BT moiety, exhibited the best performance with an EQE of 2.92% (corresponding to an LE of 4.70 cd A^{-1}) and CIE coordinates of (0.32, 0.31). It was interesting to note that higher brightness and LE could be achieved by the thermal annealing of the emissive layers at annealing temperatures from 120 to 180°C for 10 min while the EL spectra remained almost unchanged. The best performance with an EQE of 3.84% (corresponding to an LE of 6.20 cd A^{-1}) and CIE coordinates of (0.35, 0.34) was achieved after annealing for 10 min at 150°C. Besides, copolymers with similar chemical structures were also successfully synthesized by Shu and coworkers [44], and it was reported that efficient WPLEDs prepared using these copolymers exhibited an LE as high as 4.87 cd A^{-1} with CIE coordinates of (0.37, 0.36). Moreover, based on the prototype of concurrently incorporating green- and red-emitting unit, Wang et al. reported siloles containing green and red chromophores into the backbone of polyfluorene (**P25**) [45]. WPLEDs with balanced RGB emission and moderate efficiency (i.e., an LE of 2.0 cd A^{-1}) were achieved.

As mentioned above, for white light emission based on complementary colors, the incorporation of chromophores into polymer side chain is an effective approach for developing highly efficient monochromatic or full-color emission polymers. Based on this strategy, Wang's group reported that by incorporating 4-diphenylamino-1,8-naphthalimide (DPAN) and 4-(5-(4-(diphenylamino)-phenyl)-thienyl-2-)-7-(4-methoxybenzene)-2,1,3-benzothiadiazole unit (MB-BTThTPA) as green- and red-emitting units, respectively, to the side chain of polyfluorene with alkyl spacers (**P26**) [46], efficient WPLEDs with an LE as high as 7.3 cd A^{-1}, a PE of 4.2 lm W^{-1}, and CIE coordinates of (0.31, 0.32) could be obtained. In contrast, the control WLEPs (**P27**) with the green-emitting unit incorporated in the side chain and the red-emitting unit attached in the main chain of the polyfluorene only showed a much lower LE of 3.8 cd A^{-1} and PE of 2.0 lm W^{-1}. These results again indicate that attaching narrow-bandgap unit into side chain might have advantages over its incorporation into polymer backbone. Similar to the previous report [35], a WLEP was synthesized by concurrently attaching highly efficient sky-blue-, green-, and red-emitting units into the polymer side chain by alkyl spacer (**P28**) [47]. Device based on the obtained copolymer showed a very broad EL emission covering the entire visible range from 400 to 700 nm with four emission peaks, which could be attributed to the emission of the RGB chromophores and the host. As a result of this broad emission, high-quality white emission characterized by high CRI between 88 and 93 was obtained. Optimized single layer device showed a high LE of 8.6 cd A^{-1}, a PE of 5.4 lm W^{-1}, with CIE coordinates of (0.33, 0.36). More importantly, this approach for the design of these polymers does not require the polymer host to emit blue light efficiently and therefore can be benefited from the ready availability of the polymer host. It is expected to realize WPLEDs with long operational lifetime in the future.

From synthetic chemistry aspects, in addition to incorporating chromophores into polymer main chain or side chain, they can also be attached to the end group of the polymer. For instance, a novel WLEP was developed based on poly(fluorene-*co*-benzothiadiazole) backbone end-capped with a green-emitting dye *N*-phenyl-1,8-naphthalimide (**P29**) [48]. By changing the molar ratio of the comonomers, the EL spectra could be tuned for achieving white emission. The highest brightness achieved in such device configuration was 251 cd m^{-2} at a current density of 40 mA cm^{-2} with CIE coordinates of (0.31, 0.39). Scheme 3 shows the chemical structures of RGB primary colors based WLEPs utilizing singlet excitons.

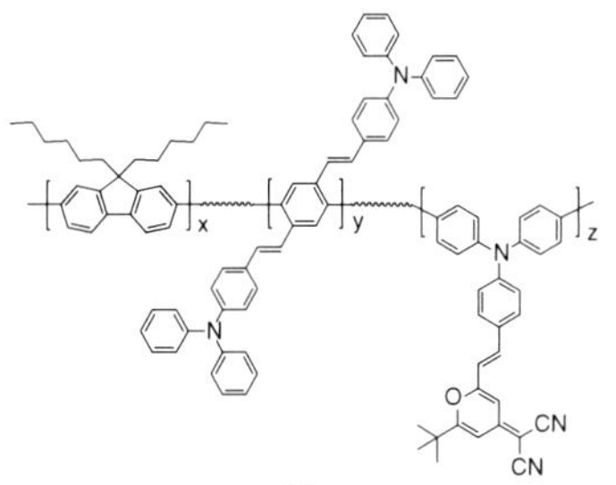

P22[41]

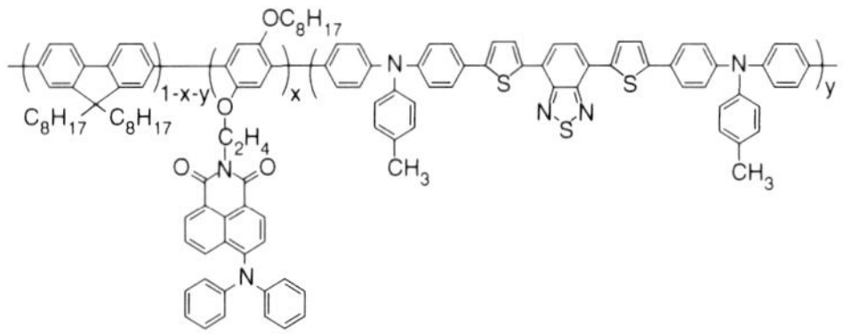

P23[42]

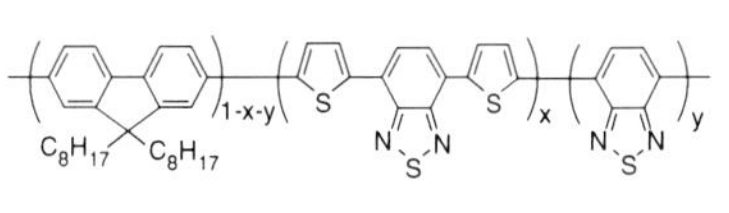

P24[43]

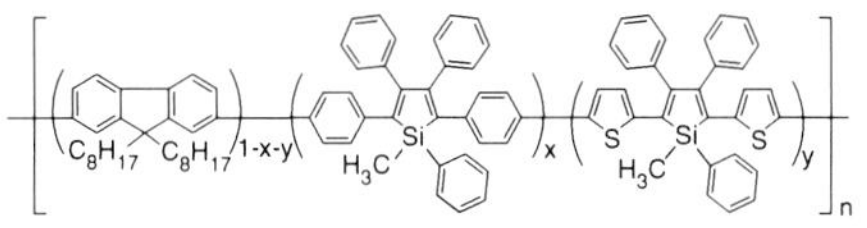

P25[45]

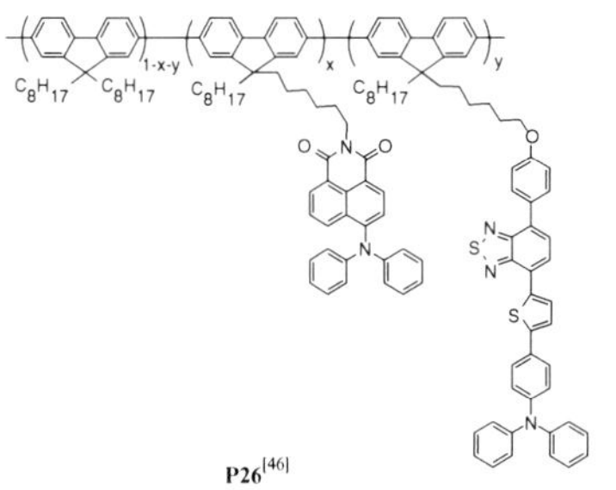

P26[46]

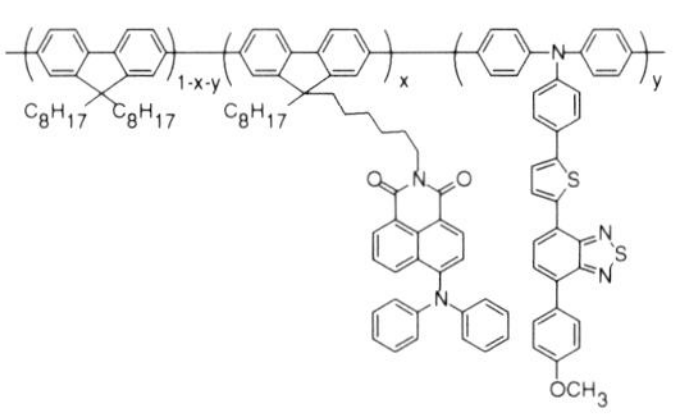

P27[46]

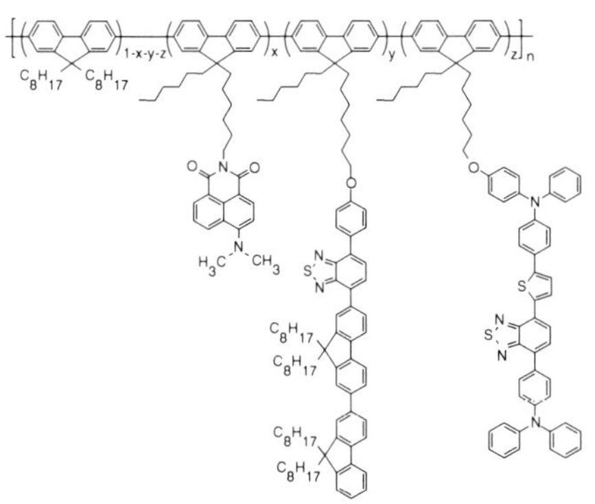

P28[47]

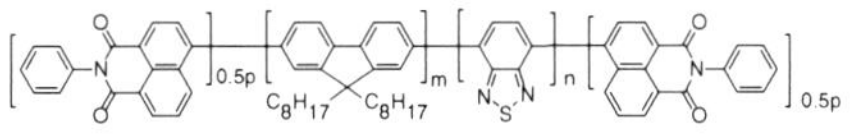

P29[48]

Scheme 3 Chemical structures of RGB primary colors based WLEPs utilizing singlet excitons

3.2.2 RGB Primary Colors Based WLEPs Utilizing both Singlet and Triplet Excitons

As discussed in Sect. 2.2, organic or polymer light-emitting devices employing phosphorescent emitters as active layer allow for a conversion of up to 100% of injected charges into emitted photons. Given this guideline, our group developed a new strategy to realize efficient white emission from a single-component polymer consisting of both phosphorescent red and fluorescent blue and green species. The synthesis was performed by incorporating a small amount of green-emitting benzothiadiazole (BT) and small amount of red-emitting iridium complex, iridium(III) bis(2-(2'-benzo[4,5-α]thienyl)-pyridinato-$N,C^{3'}$)-2,2,6,6-tetramethyl-3,5-heptanedione [(btp)$_2$Ir(tmd)] into blue polyfluorene backbone (**P30**) [49]. By adjusting the contents of BT and [(btp)$_2$Ir(tmd)] in the polymer, the EL spectrum from a single-component polymer could be adjusted to achieve white emission. The best device performance showed a peak EQE of 3.7% and a LE of 3.9 cd A^{-1} at the current density of 1.6 mA cm^{-2} with CIE coordinates of (0.33, 0.34). The maximum luminance of 4,180 cd m^{-2} was achieved at the current density of 268 mA cm^{-2} with CIE coordinates of (0.31, 0.32). More importantly, the white emission of devices from the copolymer is stable upon the change of applied voltages, representing a promising approach to prepare white-emitting single-component polymer.

Since the effective conjugation length of the triplet chromophores [(btp)$_2$Ir(tmd)] could be distinctly extended by the adjoining fluorene segments, the emission from the triplet chromophores [(btp)$_2$Ir(tmd)] is red shifted to a longer wavelength (around 650 nm) region, in which the response of human eyes is much weaker. Therefore, it is rational to expect that both LE and color quality can be substantially improved if less saturated red emitters are incorporated into polyfluorene backbone. A novel polymer with a mixed fluorescent blue-emitting polyfluorene and green-emitting benzothiadiazole (BT) and phosphorescent red-emitting iridium(III) bis(2-(-naphthalene)pyridine-$C^{2'},N$)-2,2,6,6-tetramethyl-3,5-heptanedione [(1-npy)$_2$Ir(tmd)] was synthesized (**P31**) [50]. By optimizing the components of the narrow-bandgap species, white electroluminescence was achieved with red component peaked at 625 nm. The best device performance showed an LE of 5.3 cd A^{-1} and a PE of 2.4 lm W^{-1} with CIE coordinates of (0.32, 0.34), which is very close to the equi-energy white point (0.33, 0.33). More importantly, all of the resulted devices show very good color quality with high CRI range between 84 and 89.

Recently, Lee et al. reported a new white-emitting phosphorescent polymer which contains red iridium complex in polyfluorene main chain and green chromophore as the end group of the polymer (**P32**) [51]. The copolymer was prepared from a fluorene monomer copolymerized with a red-emitting phosphorescent iridium complex, and end-capped with a green-emitting dye (N-phenyl-1,8-naphthalimide). By carefully adjusting the molar ratio of the three components, a pure white emission with CIE coordinates of (0.33, 0.34) and a maximal luminance of 300 cd m^{-2} was obtained.

Besides chelating iridium ligands conjugatively into the polymer backbone, such as **P30** and **P31**, WLEPs could also be achieved by incorporating ancillary β-diketonate ligand into the polymer main chain [52]. Zhang et al. reported the synthesis of a series of novel fluorene-based copolymers with green-emissive fluorenone defect and an iridium complex on the main chain (**P33**) by Suzuki polycondensation. By finely optimizing the contents of red phosphor and green-emitting fluorenone in polyfluorene, a series of white-emitting copolymers was successfully synthesized. It was found that single-layer PLEDs with the configuration ITO/PEDOT/copolymer/CsF/Al exhibited a maximum LE of 5.50 cd A^{-1} with CIE coordinates of (0.32, 0.45).

In parallel, a novel white electroluminescent polymer was prepared by introducing a small amount of fluorescent green-emitting benzothiadiazole (BT) unit into the blue-emitting polyfluorene backbone and covalently attaching a small amount of phosphorescent red-emitting iridium complexes, iridium(III) bis(2-(2-phenylquinoline-$N,C^{2'}$)acetylacetonate [Ir(Phq)], into the side chain of polyfluorene (**P34**) [53]. By changing the contents of the BT and/or iridium complex, the obtained EL spectra composed of fluorescent blue, green emission, and phosphorescent red emission could be readily adjusted to achieve white emission. For example, the devices fabricated from PFBT1-Phq2 and PFBT3-Phq2 emitted white light with CIE coordinates of (0.34, 0.33) and (0.32, 0.33), respectively, which are very close to standard white emission of (0.33, 0.33). In addition, the optimized device showed an LE of 6.1 cd A^{-1} with CIE coordinates of (0.32, 0.44). It is found that the white emission of devices from the copolymers is stable at wide ranges of applied voltages, and the LE declined slightly with increasing current densities. Moreover, compared with the previously reported white electroluminescent polymers derived from complementary colors containing red phosphorescent iridium and blue fluorescent polyfluorene [37], this kind of polymer showed better color purity for white emission and higher EL efficiency. This was ascribed to the exciton confinement of the incorporated BT unit and the reduced excited energy loss during the energy transfer or the prevention of the excitons from migrating to quenching sites.

Similar to the aforementioned work, Wu et al. reported the synthesis of an efficient WLEP by covalently attaching a green fluorophore 2,1,3-benzothiadiazole (BT) and a red phosphor bis[1-phenylisoquinolinato-C^2,N]iridium(III) picolinate [Ir(piq)] into the backbone and the side chains, respectively (**P35**) [54]. In the meantime, triphenylamine (TPA) and oxadiazole (OXD) moieties were introduced as pendant groups of polyfluorene to lower the injection barrier height from the electrodes and improve the transport properties. As a result of perfect charge transport in the WPLEDs based on the obtained copolymer, a low turn-on voltage of 2.8 V was observed, effectively maintaining low power consumption and obtaining high PE in these devices. By carefully controlling the concentrations of the red- and green-emitting species in the resulting polymers and by utilizing an efficient electron-injection cathode, white emission was achieved by the balanced emission of the three primary colors with a FWHM of 205 nm, a peak LE of 8.2 cd A^{-1}, a peak PE of 7.2 lm W^{-1}, and CIE coordinates of (0.35, 0.38). In addition, high CRI of 82 was recorded at all of the measured current densities. Scheme 4 shows the chemical structures of RGB primary colors based WLEPs utilizing both singlet and triplet excitons.

Apart from the extensively investigated polymers in which luminescent iridium complexes were incorporated, osmium complex containing white-emitting

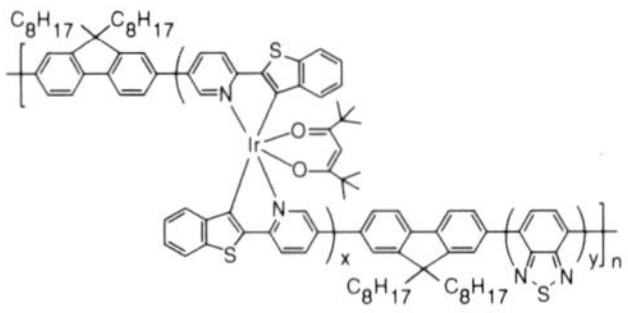

P30[49]

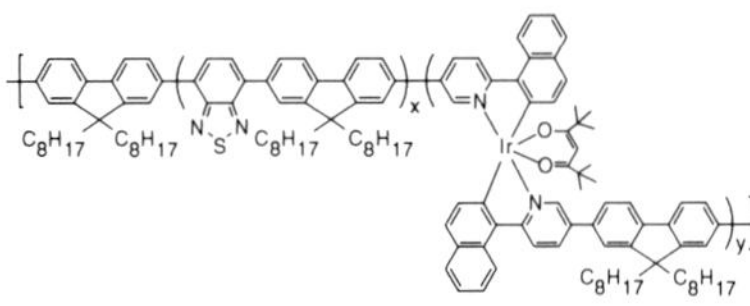

P31[50]

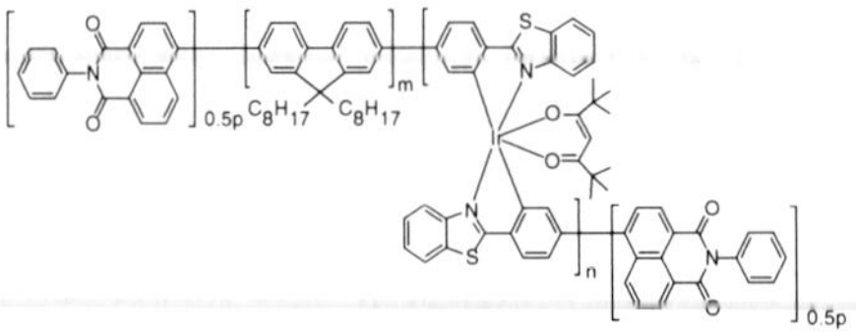

P32[51]

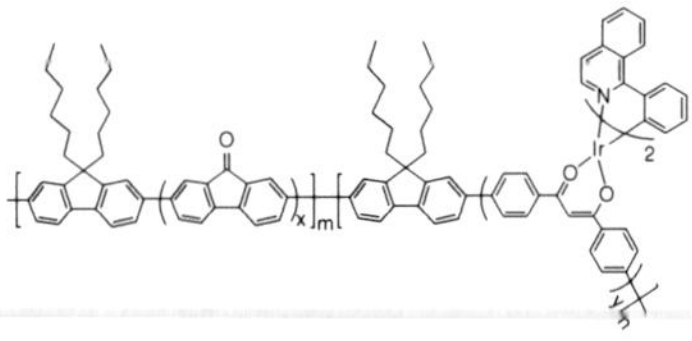

P33[52]

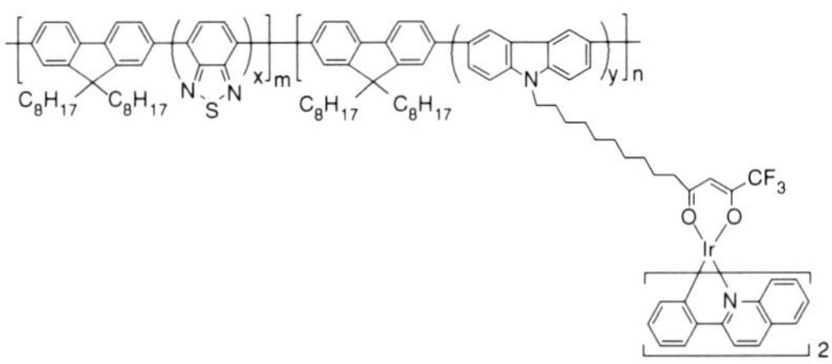

P34[53]

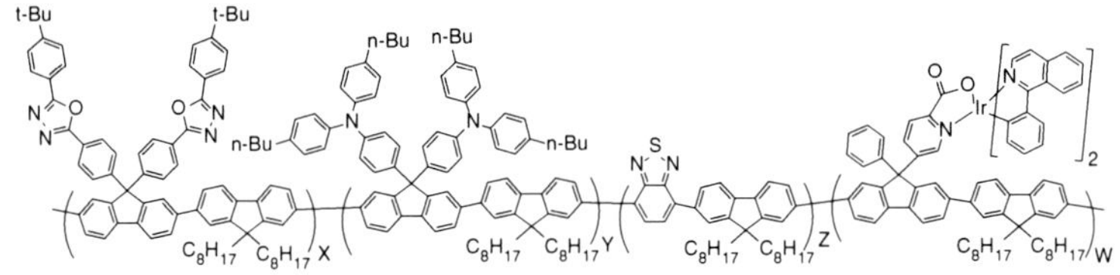

P35[54]

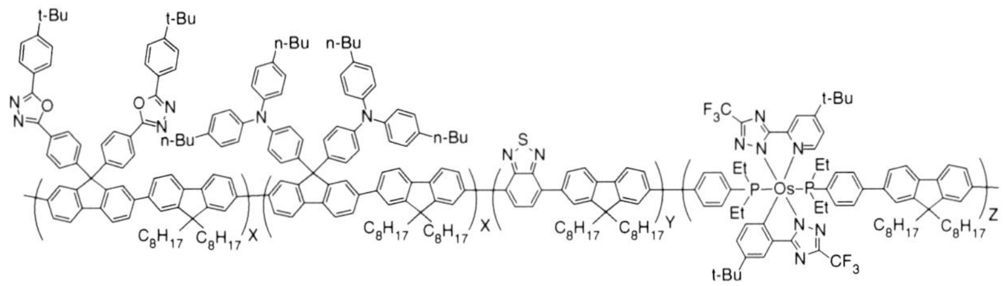

P36[55]

Scheme 4 Chemical structures of RGB primary colors based WLEPs utilizing both singlet and triplet excitons

copolymers are of special interest due to their shortened radiative lifetime and lower oxidation potential relative to the corresponding iridium-based systems. Recently, novel WLEPs incorporating red $Os(bpftz)_2(CO)_2$ complex and a certain amount of green benzothiadiazole (BT) into polyfluorene backbone were synthesized (**P36**) [55]. With device structure of ITO/PEDOT/polymer/TPBI/LiF/Al, efficient white electroluminescence displaying simultaneous blue, green, and red emissions was obtained with an LE of 10.7 cd A^{-1} (corresponding to an EQE of 5.4%) and CIE coordinates of (0.37, 0.30). It is important to note that with the luminance increasing from 32 to 8,300 cd m^{-2}, the emission color shifted from (0.36, 0.29) to (0.39, 0.32), representing a stable white emission. As a short summary of this section, device performance of WPLEDs based on single polymers is listed in Table 5 for comparison.

4 White Polymer Light-Emitting Devices

4.1 WPLEDs Using Dye-Dispersed Polymer Blend

Practically, WOLEDs are obtained by doping host materials with various dyes or narrow-bandgap polymers to achieve balanced emission. This has become a routine strategy for the fabrication of WOLEDs and has been applied for WPLEDs provided that small molecular dopants are compatible with polymer matrix and can be solution processed. From the point of view of device engineering, WPLEDs using dye-dispersed polymer blend can fully exploit the potential of low-cost fabrication of polymer optoelectronic devices.

4.1.1 WPLEDs Using Fluorescent Dye-Dispersed Polymer Blend

Fan et al. demonstrated the fabrication of efficient fluorescent WPLEDs with PFO as host, doped with orange-red-emitting small molecular chromophores 4,7-di-2-thienyl-2,1,3-benzothiadiazole (DBT, **C1** in Scheme 5), 4,7-(2'-diselenophenyl)-2,1,3-benzothiadiazole (SeBT, **C2** in Scheme 5), and 4,7-bis(N-methylpyrrol-2-yl)-2,1,3-benzothiadiazole (PBT, **C3** in Scheme 5) prepared by solution processing [56]. By carefully adjusting the dopant concentration, a nearly pure white emission with CIE coordinates of (0.33, 0.34) was obtained, and a high LE of 7.5 cd A^{-1} was recorded at luminance over 1,000 cd m^{-2}. EL spectra of the obtained devices were very stable at a wide range of operating voltage.

Xu et al. reported the fabrication of efficient WPLEDs based on blue-emitting polyfluorene doped by a "twistacene", 6, 8, 15, 17-tetraphenyl-1.18, 4.5, 9.10, 13.14-tetrabenzoheptacene (TBH, **C4** in Scheme 5) [57]. The "twistacene" of the terminal pyrene moieties not only can stabilize the inherently unstable heptacene but also can enable the oligoacene to be a strongly fluorescent molecule. The best device performance was found to be 3.55 cd A^{-1} at 4,228 cd m^{-2} with maximal PE of 1.6 lm W^{-1} at 310 cd m^{-2} and maximal luminance of 20,000 cd m^{-2}. It was found that the device gave a stable white emission via energy transfer from the blue polyfluorene to TBH by doping 1% TBH.

Huang et al. also reported highly efficient fluorescent WPLEDs based on PFO and orange-emitting dopant rubrene (**C5**, Scheme 5) [58]. It was found that by

Table 5 Summary of device performance of WPLEDs based on single-component polymer

Single-component white-emitting polymer	CIE (x, y)	λ_{max} (nm)	LE (cd A^{-1})	EQE (%)	PE (lm W^{-1})	References	Remarks
WLEPs based on blue–yellow fluorescent chromophores	0.25, 0.35	442, 560	5.3	–	2.8	[25]	–
	0.28, 0.24	436, 572	4.1	2.6	2.1	[26]	–
	0.31, 0.28	436, 572	3.5	2.3	1.7	[26]	–
	0.33, 0.31	423, 450, 479, 580	0.6	–	–	[27]	–
	0.32, 0.30	424, 449, 486, 516, 573, 623	2.0	–	–	[28]	–
	0.27, 0.35	425, 445, 540, 580	3.5	1.3	2.2	[29]	–
	0.30, 0.36	–	4.1	–	–	[30]	–
	0.41, 0.37	495, 650	1.2	1.1	–	[31]	–
	0.32, 0.36	452, 567	3.8	1.5	2.0	[32]	–
	0.35, 0.32	433, 465, 574	9.0	3.8	5.8	[33]	–
	0.30, 0.40	432, 460, 545	10.7	–	6.7	[34]	–
	0.31, 0.36	470, 568	12.8	5.4	8.5	[35]	–
	0.35, 0.39	420, 440, 562	7.1	–	4.4	[36]	–
WLEPs based on mixed fluorescent/phosphorescent blue–yellow chromophores	0.33, 0.32	440, 580	3.2	–	–	[37]	–
	0.44, 0.32	451, 591	4.5	–	2.4	[38]	–
	0.33, 0.50	–	4.6	–	–	[39]	–
WLEPs based on RGB fluorescent chromophores	0.33, 0.35	–	0.1	–	–	[41]	–
	0.31, 0.34	445, 515, 624	1.6	–	0.8	[42]	–
	0.37, 0.36	427, 452, 519, 601	4.9	2.2	–	[44]	–
	0.35, 0.34	438, 518, 616	6.2	3.8	–	[43]	–
	0.33, 0.36	450, 505, 574	2.0	–	–	[45]	–
	0.31, 0.32	436, 460, 496, 570	7.3	–	4.2	[46]	–
	0.33, 0.36	421, 475, 508, 593	8.6	–	5.4	[47]	CRI = 88
	0.31, 0.39	430, 460, 500, 580	0.7	–	–	[48]	–
WLEPs based on mixed fluorescent/phosphorescent RGB chromophores	0.33, 0.34	420, 450, 520, 660	3.9	3.7	–	[49]	–
	0.34, 0.36	420, 440, 520, 625	5.3	2.7	–	[50]	CRI = 88, CCT = 5,094 K
	0.32, 0.45	–	5.5	2.2	–	[52]	–
	0.32, 0.44	–	6.1	–	–	[53]	–
	0.34, 0.38	–	8.2	3.7	7.2	[54]	CRI = 82, CCT = 6,000 K
	0.37, 0.30	428, 518, 614	10.7	5.4	–	[55]	–

White-Emitting Polymers and Devices

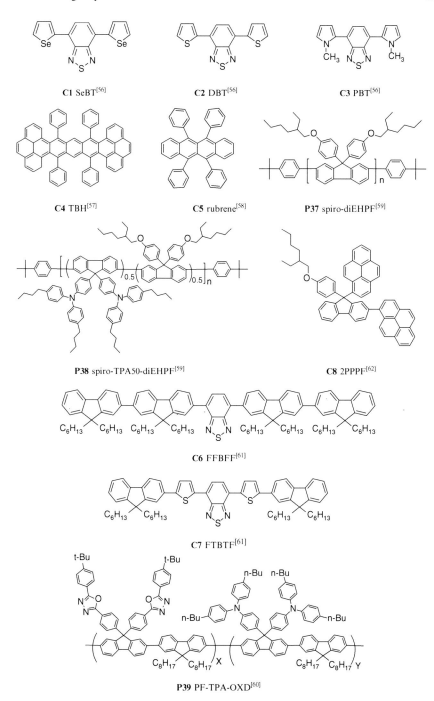

Scheme 5 Fluorescent dyes for WPLEDs

adding an electron-transporting material 2-(4-biphenylyl)-5-(4-*tert*- butylphenyl)-1,3,4-oxadiazole (PBD) into the PF-rubrene system, a more balanced charge transport was achieved, resulting in a doubled device efficiency. Furthermore, these devices could easily reach high luminance at low driving voltages, thus achieving high PE at high luminance (14.8 lm W^{-1}, 13.5 lm W^{-1}, and 12.0 lm W^{-1} at the luminances of 1,000 cd m^{-2}, 2,000 cd m^{-2}, and 4,000 cd m^{-2}, respectively). By optimizing the blend concentration of PBD, a peak LE of 17.9 cd A^{-1} and a peak PE of 16.3 lm W^{-1} were realized with CIE coordinates of (0.33, 0.43) at 25 mA cm^{-2}. The reported PE was comparable to that of incandescent lighting bulb and is among one of the highest values reported in WPLEDs.

Peng et al. reported WPLEDs with extremely stable emission color from ambipolar polyspirofluorenes doped with rubrene [59]. The polyspirofluorenes used were spirodiethylhexyloxylpolyfluorene (spiro-diEHPF, **P37**, Scheme 5) and triphenylamine-grafted spirodiethylhexyloxylpolyfluorene (spiro-TPA50-diEHPF, **P38**, Scheme 5), which showed a high charge carrier mobility in the order of 10^{-3}– 10^{-4} $cm^2V^{-1}s^{-1}$ that can serve as good host for fluorescent dye doping. Such WPLEDs based on spiro-diEHPF exhibited a maximal LE of 3.5 cd A^{-1}, a maximal luminance of 36,000 cd m^{-2}, and CIE coordinates of (0.32, 0.35), while an enhanced performance with a maximal LE of 9.0 cd A^{-1}, a maximal luminance of 36,000 cd m^{-2}, and CIE coordinates of (0.35, 0.40) were observed for the spiro-TPA50-diEHPF-based device. The performance improvement may be due to the presence of TPA groups and subsequently the enhanced hole injection.

The modification of a blue-emitting polyfluorene-derived copolymer (poly[(9,9-bis(4-di(4-n-butylphenyl)aminophenyl))]-stat-(9,9-bis(4-(5-(4-*tert*-butylphenyl)-2-oxadiazolyl)-phenyl))-stat-(9,9-di-n-octyl)fluorene) (PF-TPA-OXD, **P39**) [60] was also demonstrated, in which a series of WLEPs were prepared by doping green-emitting (4,7-bis-(9,9,9′,9′-tetrahexyl-9H,9′H-[2,2′]bifluoren-7-1)-benzo[1,2,5]thiadiazole) (FFBFF, **C6** in Scheme 5) and red-emitting (4,7-bis-[5-(9,9-dihexyl-9H-fluoren-2-yl)-thiophen-2-yl]-benzo[1,2,5]thiadiazole) (FTBTF, **C7** in Scheme 5) into polymer PF-TPA-OXD (**P39**) [61]. Such WPLEDs reached a maximum EQE of 0.82% and maximum brightness of 12,900 cd m^{-2} at 12 V, and more importantly, CIE coordinates of the device only show slight changes over a wide operating voltage ranging from 6 (0.36, 0.37) to 12 V (0.34, 0.34).

In order to prepare WPLEDs based on dye-dispersed polymer blend, the polymer solution must be capable to be spin processed. In other words, the dye must be able to be completely dissolved in the polymer solution. Recently, Liu et al. employed a novel small molecule 9-(4′-(2″-ethylhexyloxyphenyl))-2,9-dipyrenylfluorene (2PPPF, **C8** in Scheme 5) as host for orange-emitting polymer MEH-PPV [62]. The pyrene-functioned diarylfluorenes 2PPPF showed a strong aggregated emission with a broad structureless peak at about 460 nm in solid thin film. Single-layered white-emitting device by using 2PPPF:MEH-PPV blend as the active emitter demonstrated an LE up to 1.84 cd A^{-1} and brightness over 5,700 cd m^{-2}.

4.1.2 WPLEDs Using Phosphorescent Dye-Dispersed Polymer Blend

In general, WPLEDs based on phosphorescent dye-dispersed polymer blend could have much higher efficiencies than WPLEDs based on all fluorescent emitters due to the merit of 100% IQE. Up to now, WPLEDs utilizing phosphorescent dyes had drawn intense attention from many groups. Attar et al. reported efficient WPLEDs based on a single active layer containing blue-emitting poly(9,9-bis(2-ethylhexyld-fluorene-2,7-diyl) end-capped with bis(4-methyl)phenyldphenylamine; (PF2/6 am4), and yellow-orange-emitting iridium [tri-fluorenyl]pyridine complex [Ir(Fl$_3$Py)$_3$] (**C9**, Scheme 6) [63]. The fluorene-like ligands in the blended device prevented phase segregation and also enhanced energy transfer from the polymer host to the guest due to efficient overlap of wave function (Dexter process) and host singlet emission and guest absorption bands (Förster process) that reduced the loading level required to produce white emission. A stabilized white emission was obtained when the doping concentration was fixed at 2–3 wt%. The optimized WPLEDs showed CIE coordinates of (0.348, 0.367) with a peak EQE of 2.8% and a peak LE of 4.6 cd A^{-1}.

In order to improve the color quality of WPLEDs, Niu et al. reported WPLEDs with CRI as high as 92 by doping a WLEP with a red phosphorescent dopant bis(1-phenylisoquinolyl)iridium(III)(1-trifluoro)acetylacetonate [(Ppq)$_2$Ir(acac)] (**C10**, Scheme 6) [64]. Compared with the device based on the single white emission polymer, the Ir-doped WPLEDs showed a broader emission spectrum and higher efficiencies. By controlling the contents of the doped electron-transporting PBD and the red phosphorescent dopant, white emission with CIE coordinates of (0.34, 0.35) and an LE as high as 5.3 cd A^{-1} were obtained.

Gong and coworkers reported efficient WPLEDs with single active layer based on PFO doped with an orange triplet emitter, tris(2,5-bis-2′-(9′,9′-dihexylfluorene) pyridine)iridium(III) [Ir(HFP)$_3$] (**C11**, Scheme 6) and green-emitting poly(9,9-dioctylfluorence-co-fluorenone) with 1 mol% fluoreneone (PFO-F (1 mol%)) [65]. By finely optimizing the blend ratio, a series of WPLEDs was fabricated with an LE between 3.0 and 4.3 cd A^{-1}, CIE coordinates in the range (0.329, 0.321) to (0.352, 0.388). It is important to note that these devices were characterized by very good color quality, in terms of both CRI (86–92) and CCT (4,600–6,400 K), that is very suitable for both solid-state lighting and displays applications.

Similarly, our group demonstrated efficient WPLEDs based on a highly efficient blue emitter, polyhedral oligomeric silsesquioxane-terminated poly(9,9-dioctylfluorene) (PFO-poss) dual-doped with two phosphorescent dyes, *fac*-tris[2-(4′-*tert*-butyl) phenylpyridine]iridium(III) [Ir(Bu-ppy)$_3$] (**C12**, Scheme 6) and iridium(III)[bis-(1-phenylisoquinolyl)](1,1,1-trifluoroacetylacetonate) [(Piq)$_2$Ir(acaF)] (**C13**, see Scheme 6), as green and red emitters, respectively [66]. By adjusting the concentrations of the green and red emitters, efficient and pure white light emission with CIE coordinates of (0.33, 0.33) was obtained at a luminance of 1,240 cd m^{-2}. The peak LE and PE reached 9.0 cd A^{-1} and 5.5 lm W^{-1}, respectively.

Shih et al. reported WPLEDs by the doping of a blue fluorescent dye, 4,4′-bis[2-{4-(*N,N*-diphenylamino)phenyl}vinyl]biphenyl (DPAVBi) and a phosphorescent

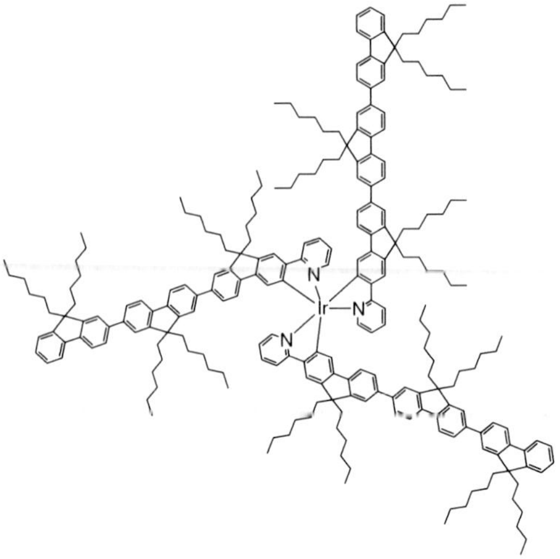

C9 Ir(FlPy)$_3$[63]

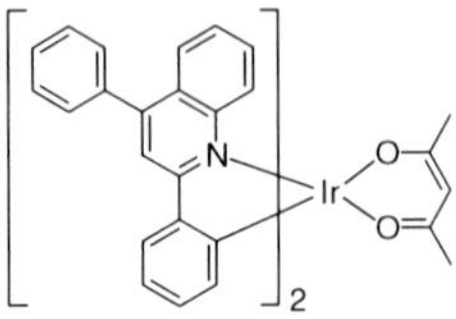

C10 (Ppq)$_2$Ir(acac)[64]

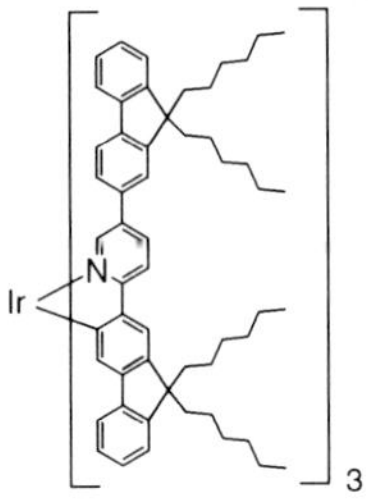

C11 Ir(HFP)$_3$[65]

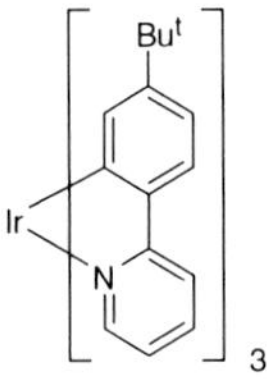

C12 Ir(Bu-ppy)$_3$[66]

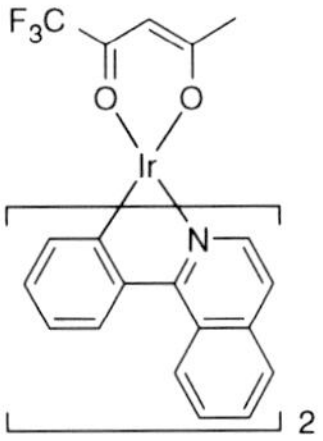

C13 (Piq)$_2$Ir(acaF)

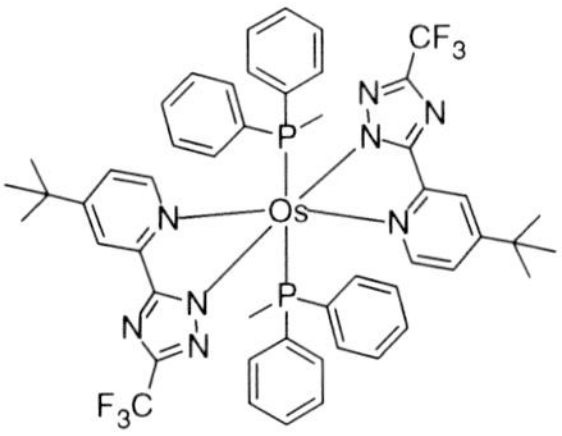

C14 Os(bpftz)[67]

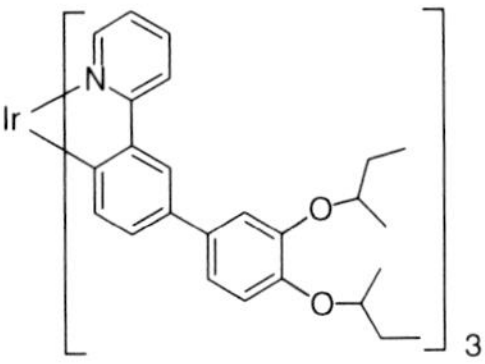

C15 Ir(PBPP)$_3$[68]

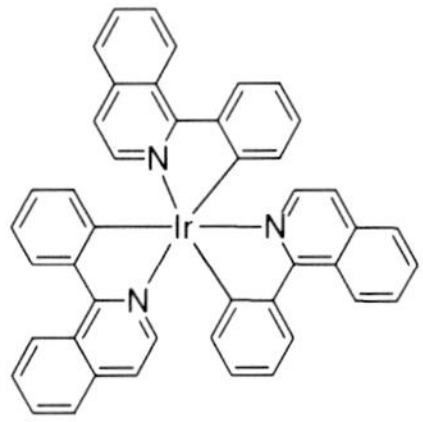

C16 Ir(PIQ)$_3$[68]

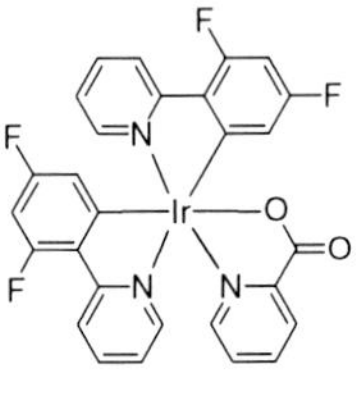

C17 FIrpic[22, 69]

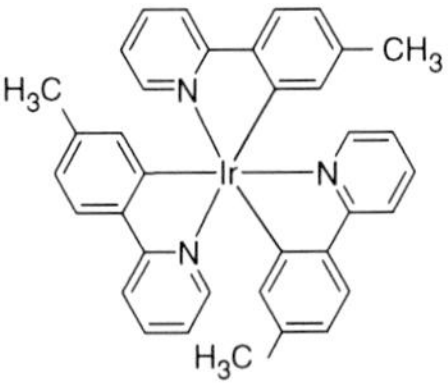

C18 Ir(mppy)$_3$[22]

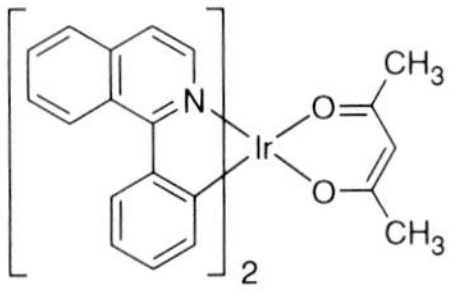

C19 Ir(piq)[22]

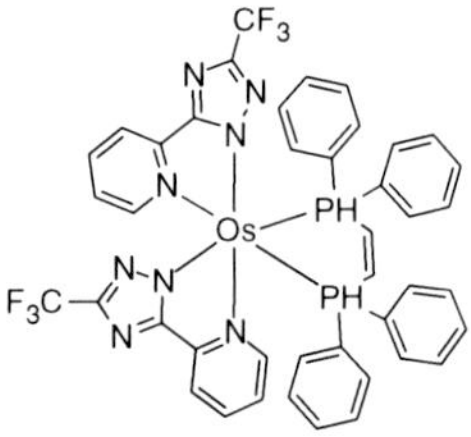

C20 Os(fptz)$_2$(dppe)[69]

Scheme 6 Phosphorescent dyes for WPLEDs

orange-emitting osmium complex [Os(bpftz)] (**C14**, Scheme 6), into a PVK host blended with 30 wt% of electron-transporting PBD [67]. This dual-doped device exhibited an intense white emission with CIE coordinates of (0.33, 0.34) and a high LE of 13.2 cd A^{-1}. In addition, the color remained unchanged even at a high luminance of 10,000 cd m^{-2}.

Kim et al. developed efficient WPLEDs using a polyfluorene-type blue-emitting conjugated polymer doped with a green phosphorescent emitter [Ir(PBPP)$_3$] (**C15**, Scheme 6) and a red phosphorescent emitter [Ir(piq)$_3$] (**C16**, Scheme 6) [68]. In order to improve the miscibility between conjugated polymer and iridium complexes due to their poor chemical compatibility and phase separation, PVK was selected and blended with the conjugated polymer to form the host. By controlling the concentration of dye and the phase morphology of the blended film, nearly "pure" white light was realized with CIE coordinates lying in between (0.34, 0.34) and (0.32, 0.32). The recorded peak LE of the resulting devices was 12.52 cd A^{-1} at a current density of 7.23 mA cm^{-2}, driving voltage of 6.2 V, and a brightness of 905 cd m^{-2}, corresponding to an EQE of 3.2%. As a result of simultaneous red, green, and blue emission covering the entire visible spectrum, high CRI of about 83–86 was obtained, representing good color quality which is appropriate for lighting applications.

Nowadays, in order to further improve the efficiencies of WOLEDs, the most exploited and effective approach has focused on the system employing fluorescent host and RGB phosphorescent dopants [19]. Such kind of WOLEDs can allow for a conversion of up to 100% of injected charges (both singlet and triplet excitons can be harvested) into emitted photons, leading to a theoretical IQE of unity. This strategy had been applied to the fabrication of highly efficient polymer-based white-emitting devices [22, 23, 69–71]. Our group reported efficient single layer WPLEDs with an emission layer containing a blend of two or three phosphorescent iridium complexes within a PVK/1,3-bis[(4-*tert*-butylphenyl)-1,3,4-oxadiazolyl] phenylene (OXD-7) host matrix [22]. We relied on iridium(III) bis(2-(4,6-difluorophenyl)-pyridinato-$N,C^{2'}$) picolinate ([FIrpic], **C17**, sky-blue emission), iridium(III) tris(2-(4-tolyl)pyridinato-$N,C^{2'}$) ([Ir(mppy)$_3$], **C18**, green emission), and iridium(III) bis(1-phenylisoquinoline) (acetylacetonate) ([Ir(piq)], **C19**, red emission). Scheme 6 shows their molecular structures. Particularly, a series of devices with various dopant concentrations and ratios was fabricated and evaluated. It was found that, at forward viewing, the triple-doped WPLEDs had a peak PE of 10.0 lm W^{-1} at 6.9 V and a peak LE of 24.3 cd A^{-1} at 20.8 mA cm^{-2}, while for the dual-doped device, a peak PE of 6.4 lm W^{-1} at 7.5 V and a peak LE of 16.1 cd A^{-1} at 15.3 mA cm^{-2}. In addition, CIE coordinates of (0.34, 0.47) and (0.33, 0.36) were realized for triple-doped and dual-doped devices at a current density of 12 mA cm^{-2}, respectively. Such devices also demonstrated an appropriate CCT within 2,500–6,500 K and a relative higher CRI of up to 77, which were suitable for commercial applications as solid-state lighting.

It is important to note that the turn-on voltage (defined as the voltage where 1 cd m^{-2} is measured) of 3.9 V is quite low among PVK-based white PLEDs, where usually high drive voltages are required because of low hole transport mobility (~10^{-5} m^2 V^{-1} s^{-1}), unmatched charge carrier-injection barrier heights, and charge-trapping effects. The low operation voltages of the devices are important for reducing power consumption. The simplicity of the device structure and the fabrication method, together with the high efficiencies observed, make these white-emitting PLEDs promising candidates for lighting applications.

4.2 WPLEDs from All-Polymer Blends

Apart from single-component polymer consisting of different chromophores and host–dopant system, white emission can be realized from polymer blends in which appropriate ratios of red, green, and blue phosphorescent or fluorescent dopants are mixed. The major advantage of this approach lies in its very simple process, and the emission color can be tuned readily.

4.2.1 WPLEDs from Fluorescent Polymer Blends

Shu et al. reported WPLEDs formed from polymer blends consisting of blue and orange polyfluorene copolymers [60]. Of which, PF-TPA-OXD (**P39**, Scheme 5) was used as efficient blue emitter while PFTO-BSeD5 (with molecular structure similar to **P6** [30]), a new orange-emitting polymer that incorporated 5 mol% of narrow-bandgap benzoselenadiazole (BSeD), units into the polyfluorene backbone for yellow emitter. Since both the host and dopant polymers possess hole-transporting triphenylamine moieties (TPA units) and electron-transporting oxadiazole moieties (OXD units) in their side chain, it is expected that balanced charge transport can be obtained in the resulting devices. The obtained device displayed CIE coordinates of (0.32, 0.33) with a high LE up to 4.1 cd A^{-1}. In addition, phase separation between the host and dopant could be suppressed effectively because of the very similar chemical structures of these two polymers, leading to very stable emission color over a relatively wide range of operating bias (from 11 to 21 V).

Recently, Hsieh et al. reported two new orange or yellow chromophores with phenothiazine and thiophene units as the central structure of the chromophores incorporating into polyfluorene main chain with content of 1 mol% (PFPhT2, **P40**; PFThT2, **P41**) [72]. The cyano group is well known for its electron-withdrawing characteristic which is beneficial to enhancing electron affinity of the chromophores. In addition, introducing phenothiazine or its derivatives can improve hole-transporting ability, and an enhancement on the EL efficiency could be expected. It was shown that white emission devices can be obtained by blending of the resulted copolymers with poly(9,9-dihexylfluorene), and the best device performance was an LE of 1.85 cd A^{-1} and CIE coordinates of (0.31, 0.33) for PFPhT2 (**P40**) and 0.87 cd A^{-1} for PFThT2 (**P41**).

Niu et al. also reported efficient WPLEDs based on fluorescent polymer blend in a single active layer structure, in which a copolymer of 9,9-dioctylfluorene and 4,7-di(4-hexylthien-2-yl)-2,1,3-benzothiadiazole (PFO-DHTBT), poly[2-(4-(3′,7′-dimethyloctyloxy)-phenyl)-*p*-phenylenevinylene] (P-PPV), and poly(9,9-dioctylfluorene) (PFO) were used as RGB emitters, respectively [73]. Optimized white emission with CIE coordinates of (0.31, 0.37) was achieved when the blend ratio (by weight) between PFO:P-PPV:PFO-DHTBT was fixed at 96:4:0.4, with a peak LE of 7.6 cd A^{-1}. Furthermore, it was found that the emission color was very stable upon change of applied voltages.

So far, the efficiencies of WPLEDs based on polymer blends are usually limited by the efficiency of the polyfluorenes and their derivatives (their LE are typically between 1 and 2 cd A^{-1}), and thus rarely exceed 10 cd A^{-1}. It is expected that the efficiency of all-polymer WPLEDs can be significantly enhanced if more efficient blue emitter can be used in these devices. Very recently, we reported efficient and color-stable WPLEDs based on a newly synthesized efficient blue-emitting polymer poly[(9,9-bis(4-(2-ethylhexyloxy)phenyl)fluorene-*co*-(3,7-dibenziothiene-S,S-dioxide10)] (PPF-3,7SO10, **P42**), with appropriate blend ratio with two typical electroluminescent polymers, green-emitting P-PPV and orange-red-emitting poly[2-methoxy-5-(2′-ethylhexyloxy)-1,4-phenylene vinylene] (MEH-PPV). In a single active layer WPLEDs with a blend ratio of 100:0.8:0.5 (B: G: R) by weight, white light emission with CIE coordinates of (0.34, 0.35) was observed with an LE of 14.0 cd A^{-1} (corresponding to an EQE of 6.9 %) and a PE of 7.6 lm W^{-1} [74]. The obtained LE was one of the highest reported values for WPLEDs based on all fluorescent polymer emitters. The devices had appropriate color temperature of 2,500–6,500 K and high CRI of 72–79 and were characterized by stable electroluminescent spectra upon change of current density, stress, and annealing at high temperatures. Thus, it can find application in solid-state lighting.

To obtain white-emitting devices from polymer blends, Lee et al. synthesized novel conjugated polyfluorene copolymer poly[9,9-dihexylfluorene-2,7-diyl-co-(2,5-bis(40-diphenylaminostyryl)-phenylene-1,4-diyl)] (PG10, **P43**) which can be used as a host material [75]. When blended with 1.0 wt% of a red-emitting polymer, poly[9,9-dioctylfluorene-2,7-diyl-alt-2,5-bis(2-thienyl-2-cyanovinyl)-1-(2′-ethylhexyloxy)-4-methoxybenzene-5′,5′-diyl] (PFR4-S, **P44**), or MEH-PPV. The device based on PG10: PFR4-S showed a nearly pure white emission with an LE of 0.10 cd A^{-1} and CIE coordinates of (0.33, 0.36) at 8 V; for the PG10:MEH-PPV device, the CIE coordinates at this voltage were (0.30, 0.40) and a peak LE of 0.47 cd A^{-1} was observed with a maximum brightness of 1,930 cd m^{-2}.

Tan et al. reported efficient WPLEDs based on the polymer blend of blue-emitting poly(2-(4′-(diphenylamino)phenylenevinyl)-1,4-phenylene-alt-9,9-*n*-dihexylfluorene-2,7-diyl) (PDFF, **P45**) doped with a red-emitting polymer poly{2-[3′,5′-bis(2′-ethylhexyloxy)benzyloxy]-1,4-phenylenevinylene}-*co*-poly(2-methoxy-5-(2′-ethyl-hexyloxy)-1,4-phenylenevinylene) (BE-co-MEH-PPV, **P46**) [76]. By changing the dopant concentration, white electroluminescence was achieved with a maximal LE of 0.29 cd A^{-1} and CIE coordinates of (0.33, 0.34). It was also found that the overall performance of the device was dramatically enhanced to

2.37 cd A^{-1} by introducing an Alq$_3$ layer as an electron-transporting and hole-blocking layer.

Similarly, Sun et al. reported the fabrication of WPLEDs with CIE coordinates of (0.32, 0.34) from a blend of poly(9,9-dioctylfluorene-2,7-diyl) blended with poly[5-methoxy-2-(2′-ethyl-hexylthio)-p-phenylenevinylene] (MEHT-PPV, **P45**) [77]. Using blends of poly[N,N′-bis(4-butylphenyl)-N,N′-bis(phenyl)benzidine] (poly-TPD) and poly(N-vinyl-carbazole) (PVK) as hole-transporting layers (HTLs), the LE of the devices was enhanced to 3.2 cd A^{-1}, doubling that of the control device. Scheme 7 shows the chemical structures of monochromatic fluorescent emitters for polymer blend.

4.2.2 WPLEDs from Phosphorescent Polymer Blends

Tokito et al. reported the fabrication of highly efficient WPLEDs using blue phosphorescent and red phosphorescent polymer blend (**P48, P49**) [78]. In these phosphorescent polymers, red-emitting and blue-emitting iridium complexes were chemically connected to a vinyl polymer backbone, respectively, while the carbazole unit was directly bonded to the backbone as the charge transport unit. White emission with CIE coordinates of (0.34, 0.36) was achieved when the blend ratio between the blue phosphorescent and red phosphorescent polymer was 10:1. It is important to note that the triplet energy of electron transport material has strong influence on the performance of the device based on the blue phosphorescent polymer. In order to avoid quenching of electrophosphorescence by electron-transporting material, it is a prerequisite that the electron-transporting material has a higher triplet energy or at least comparable to that of the blue Ir complex molecule. As a result, WPLEDs with 1,3-bis[(4-*tert*-butylphenyl)-1,3,4-oxadiazolyl]phenylene (OXD-7), whose triplet energy is about 2.7 eV, as electron-transporting material had an EQE of 4.5 %. In contrast, for the devices using 2,9-dimethyl-4,7-diphenyl-1,10-phenanthroline (BCP) as electron-transporting material, whose triplet energy is about 2.6 eV, a much lower efficiency was recorded.

Kappaun et al. reported that by blending polyfluorene with the phosphorescent red iridium containing polyfluorene derivatives (**P50**), white electroluminescence with CIE coordinates of (0.30, 0.35) that consisted of simultaneous blue, green, and red emission was achieved [79]. The analysis of the displayed EL spectra showed that blue emission (400–450 nm) was contributed from polyfluorene, while the green emission (500–600 nm) achieved from the keto defect of fluorenone and the incorporated iridium(III) complex contributed to the red emission band (650–700 nm).

Xiong et al. also reported polymer white-emitting devices from a blend of poly [9,9-di-(2-ethylhexyl)-fluorenyl-2,7-diyl]-end-capped with polysilsesquioxane and a chelating copolymer of poly[(9,9-bis(3-(N,N-dimethylamino)propyl)-2,7-fluorene-alt-2,7-(9,9-dioctylfluorene))-co-[2,7-(9,9-dioctlyfluorene)-alt-5,5-bis(2-(4-methyl-1-naphthalene)pyridine-$N,C^{2'}$)iridium(III) acetylacetonate]] ([PFN-NaIr], **P51**) [80].

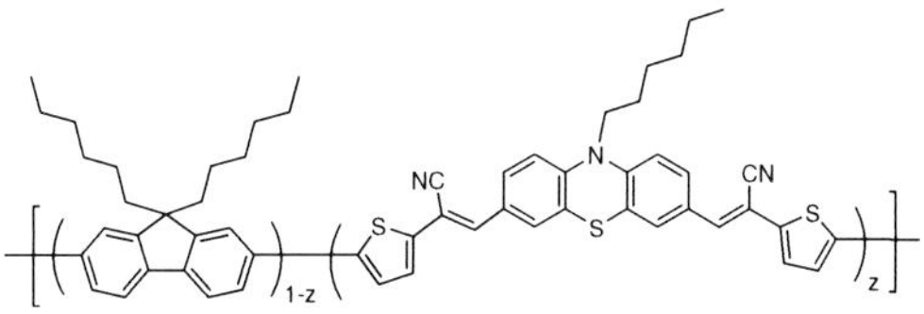

P40 PFPhT2[72]

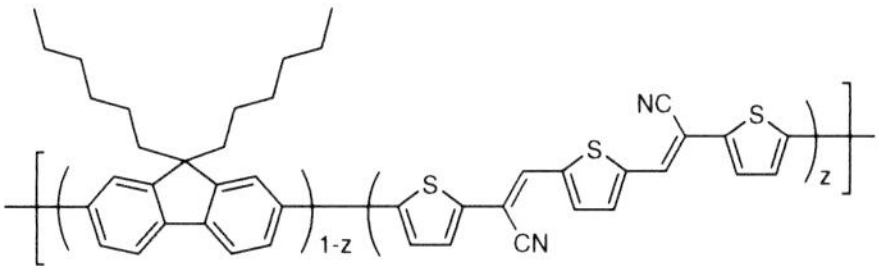

P41 PFThT2[72]

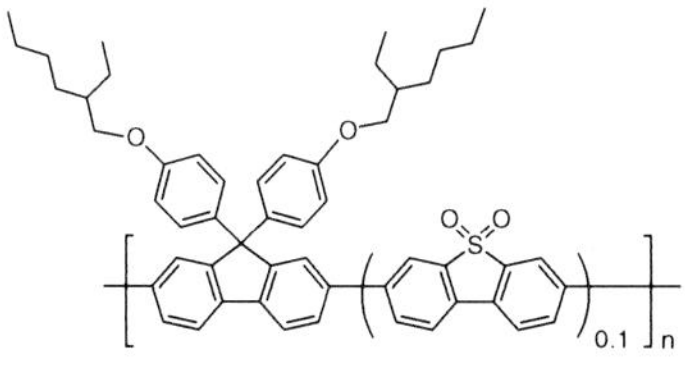

P42 PPF-3, 7SO10[74]

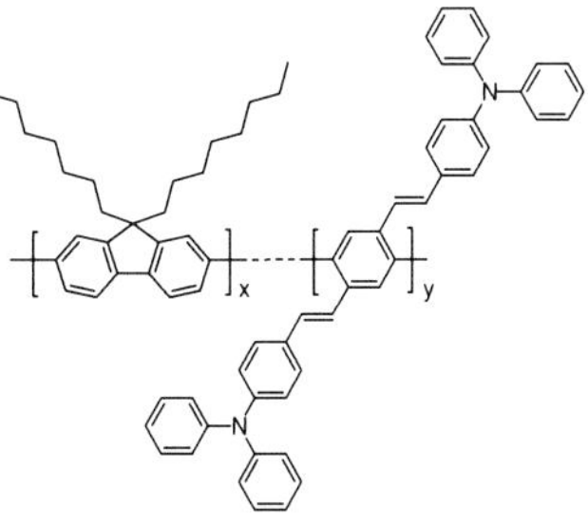

P43 PG10[75]

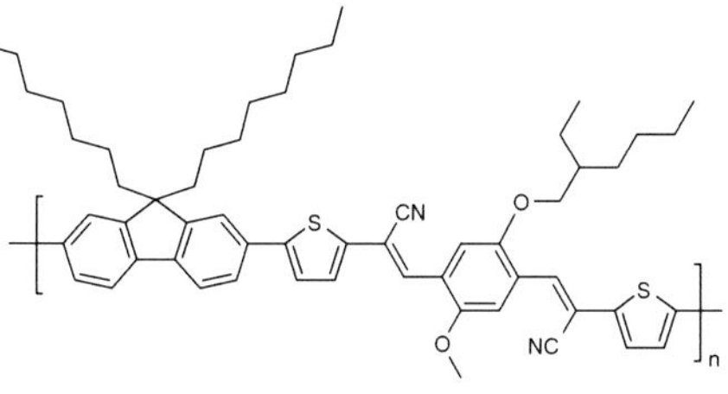

P44 PFR4-S[75]

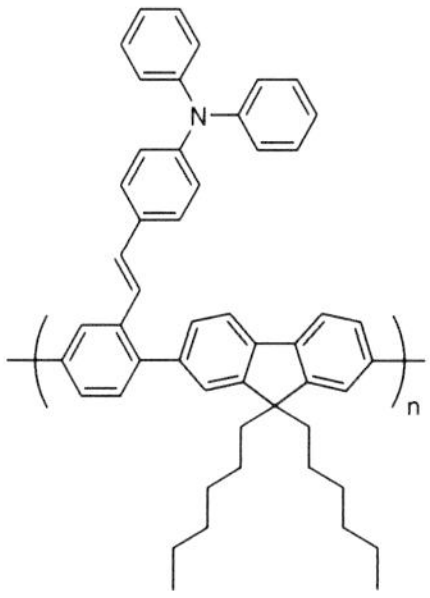

P45 PDFF[76]

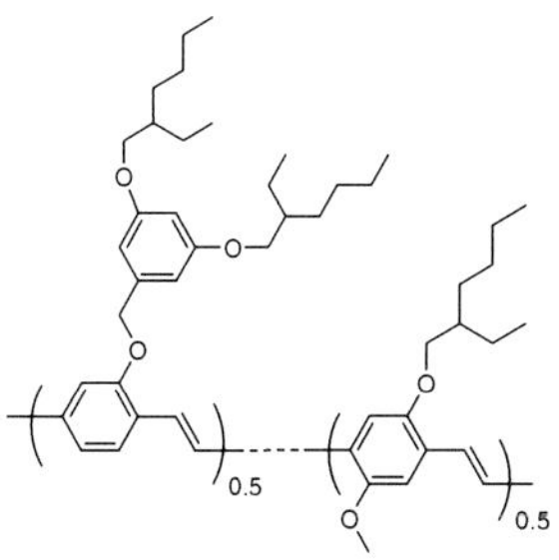

P46 BE-co-MEH-PPV[76]

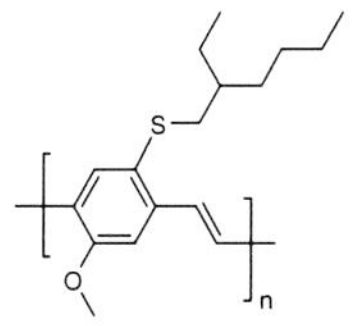

P47 MEHT-PPV[77]

Scheme 7 Monochromatic fluorescent emitters for polymer blend.

An interesting thing with the prepared white-emitting devices based on PFO:[PFN-NaIr] blend is that the plain aluminum cathode device showed comparable performance with those devices using low work function metals (such as Ba, Ca) as cathode, which is believed to be related to the interfacial dipole formed by the interaction between the polar group of the [PFN-NaIr] and Al. The white-emitting devices based on the polymer blend showed an LE of 1.31 cd A^{-1} with CIE coordinates of (0.34, 0.35). Scheme 8 shows the chemical structures of monochromatic phosphorescent emitters for polymer blend.

4.3 WPLEDs from Excimer/Exciplex

Apart from those derived from single-component polymer and polymer blend described in Sects. 4.1 and 4.2, respectively, white emission can also be obtained by utilizing excimer that occurred intermolecularly or exciplex process that formed intramolecularly.

Chao et al. reported WPLEDs with PVK and poly(2-dodecyl-*p*-phenylene) (C12O-PPP) as the two subsequent emitting layers that were capable of emitting nearly white light with a broad EL spectrum covering the range from 400 to 700 nm [81]. As a result of the mixing of the two polymers, formation of exciplex occurred by the electron in the lowest unoccupied molecular orbital (LUMO) of C12O-PPP and the hole in the highest occupied molecular orbital (HOMO) of carbazole groups in PVK, and therefore, emitting light with a broad spectrum extending to 700 nm. It was found that the EL emission of the obtained device varied with the change of temperature, which can be understood in terms of light emission via the exciplex process occurring intermolecularly, and the possibility for electrons in the LUMO of C12O-PPP to overcome the binding energy of the molecules and recombine with holes in the HOMO of PVK which has a strong dependence on temperature.

More recently, Sun et al. reported WPLEDs utilizing fluorenone defects for green emission and exciplex for red emission in ITO/poly-TPD:PVK/emissive layer/cathode [82]. The emissive layer consisted of blue-emitting poly(9,9-dihexylfluorene-*alt*-co-2,5-dioctyloxy-*para*-phenylene) (PDHFDOOP, **P52**) and green-emitting poly[6,6'-bi-(9,9'-dihexylfluorene)-co-(9,9'-dihexylfluorene-3-thiophene-5'-yl] (PFT, **P53**). By annealing the emitting layer at a relatively high temperature (up to 110°C), fluorenone defects were generated into PDHFDOOP, which subsequently formed an exciplex with poly-TPD, emitting red light. The CIE coordinates of the WPLEDs were (0.31, 0.32) at a luminance of 100 cd m^{-2}, and an LE of 3 cd A^{-1} was achieved at 10 mA cm^{-2}.

Lee et al. reported a promising oxadiazole-containing phenylenevinylene etherlinkage copolymer (**P54**), which could emit nearly white light from a single-layer device configuration [83]. The emission spectrum was composed of a red component originating from the new excited dimer in combination with a blue–green component from an individual lumophore and excimer. This excited dimer is formed under a strong electric field inside the device and cannot be produced by

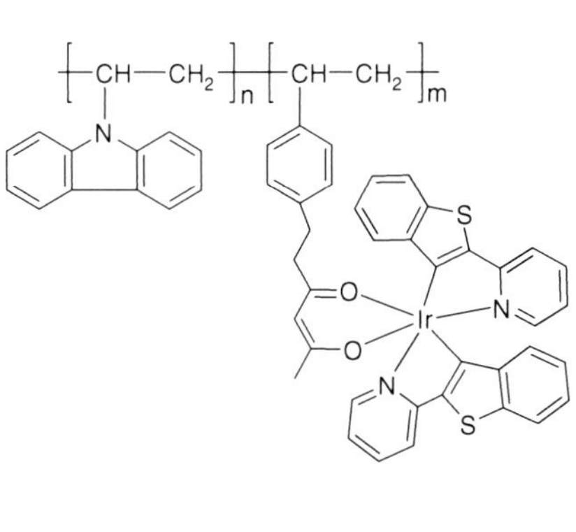

P48 RPP[78]

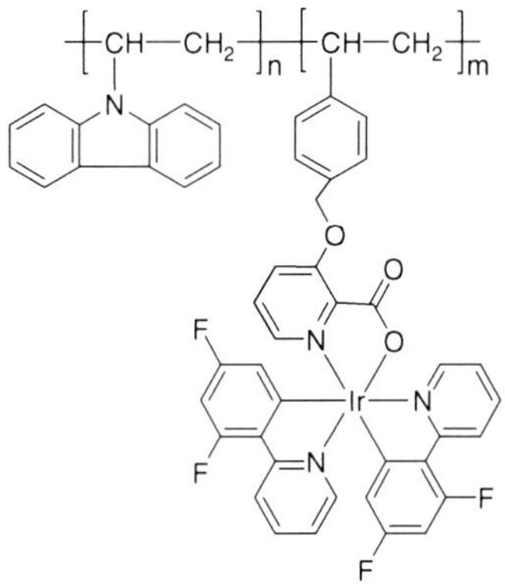

P49 BPP[78]

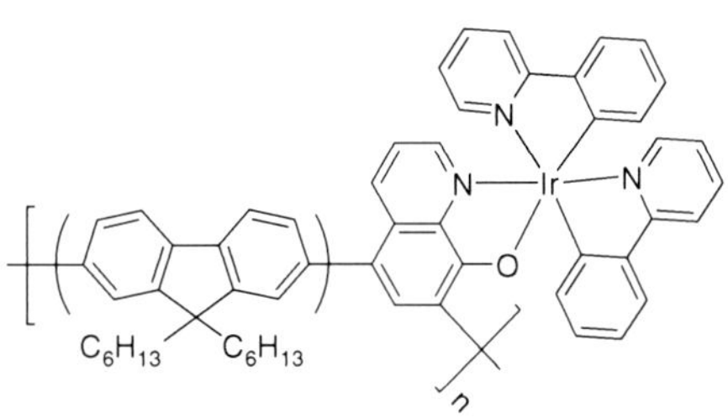

P50[79]

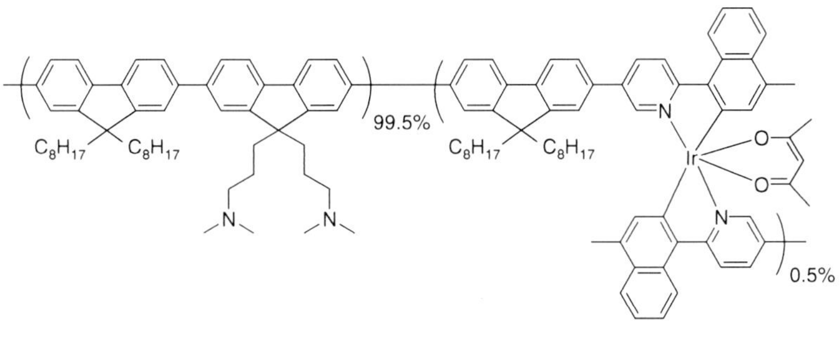

P51 PFN-NaIr05[80]

Scheme 8 Monochromatic phosphorescent emitters for polymer blend

photoexcitation, which is different from the excimer or exciplex that is often found both in photoluminescence and electroluminescence, and it is termed the "electromer". Scheme 9 shows the chemical structures of polymers with broad emission spectra.

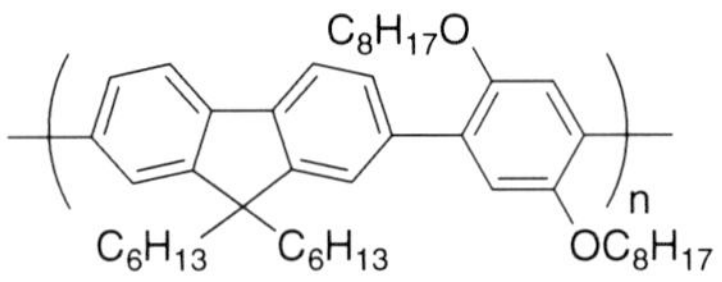

P52 PDHFDDOP[82]

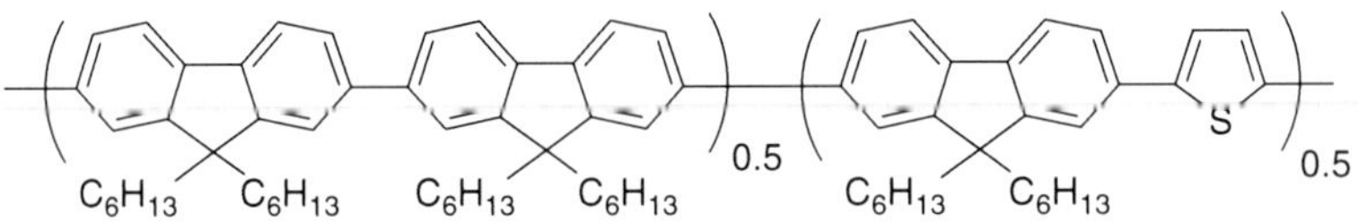

P53 PFT[83]

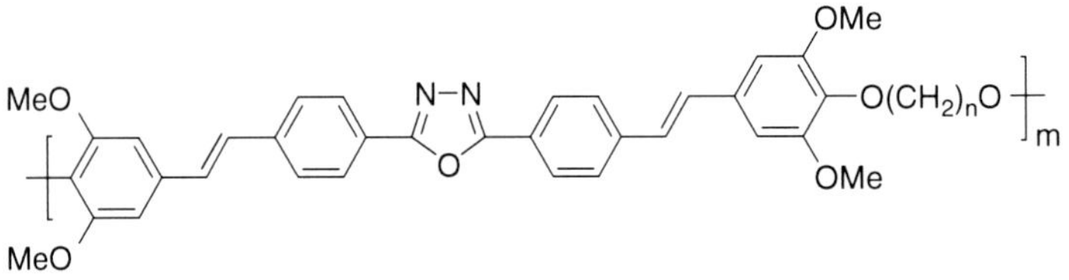

P54[83]

Scheme 9 Chemical structures for polymers with broad emission spectra

4.4 WPLEDs from Multiple Emissive Layers

Fabrication of devices with multilayer structure is a challenge for PLEDs, as mixing between each emissive layer and subsequent charge-transporting layer is a common and serious problem, and most of the organic or polymer charge-transporting materials are typically soluble in most of the organic solvents from which the EL layer is processed. In order to fabricate multilayer WPLEDs, proper choice and use of solvent is a key issue.

Our group demonstrated that efficient WPLEDs with bilayer structures can be fabricated by spin-coating process using different organic solvents to avoid interlayer mixing [84]. In these devices, the first active layer for red emission consisting of PVK and PFO-DHTBT was cast from chlorobenzene solution, while the second active layer, in which blend of polyhedral oligomeric silsesquioxane-terminated poly(9,9-dioctylfluorene) (PFO-poss) and P-PPV was incorporated for blue and green emission, was cast from *p*-xylene solution. In this device configuration, a HTL PVK was used as host for the red emitter. More importantly, PVK can resist the erosion of common nonpolar solvents from which the second layer was cast to a certain extent; hence, the dissolution of the first active layer lying beneath the second layer can be avoided. By adjusting the blend ratio and layer thickness of each active layer, pure white emission with CIE coordinates of (0.32, 0.33) and an LE of 4.4 cd A^{-1} were achieved. Zhou et al. reported double-layer-structured WPLEDs based on three primary RGB luminescent polymers [85]. The first active layer was formed by deposition of PVK and poly-(N,N'-bis(4-butylphenyl)-N,N'-bis (phenyl)benzidine (poly-TPD) binary-host in a blend ratio of 1:1 doped with poly {2-[3′,5′-bis(2″-ethylhexyloxy)benzyloxy]-1,4-phenylenevinylene}-co-poly(2-methoxy-5-(2′- ethyl-hexyloxy)-1,4-phenylenevinylene) (BE-co-MEH-PPV, **P46**) from chlorobenzene solution. After proper annealing and cooling down, the second active layer was spin-coated on top from toluene solution, in which PFO and a green-emitting polyfluorene derivative PFT (**P53**) were mixed in appropriate ratio. Based on this configuration, the resulting WPLEDs showed stable white light emission with a peak LE of 4.4 cd A^{-1} and a maximum brightness of 4,420 cd m^{-2} at 17 V, with CIE coordinates of (0.32, 0.34) at 55 mA cm^{-2}.

Moreover, Sun et al. demonstrated the fabrication of multilayer WPLEDs via a sequential solution-processing approach, in which a thin film of deoxyribonucleic acid–cetyltrimetylammonium (DNA–CTMA) complex was used as a hole-transporting/electron-blocking layer [86]. With PFO/MEH-PPV as emissive layer, the obtained WPLEDs showed an LE of 10.5 cd A^{-1} and improved color stability.

4.5 WPLEDs with Functional Charge-Injection/Transport Layers

Through over a decade of development, PLEDs have evolved from its simple sandwich prototype toward multilayer structure [87], in which each layer plays specific function involving hole transporting, electron transporting, carrier blocking, and recombination zone control. In order to achieve highly efficient PLEDs, carefully designed metal–organic interface and charge transport layer are required to achieve balanced charge carrier-injection/transport and shift of exciton recombination zone away from quenching electrode. These basic principles can also be applied in fabrication of highly efficient WPLEDs in a multilayer configuration, of which charge-injection/transport layers by solution-processed technology is of special interest since it can further ensure fully exploiting the potential of low-cost fabrication of polymer optoelectronic device.

On the basis of the previous work [65], Gong et al. further reported efficient, multilayer WPLEDs using water-soluble polyelectrolytes as the HTL and electron-transporting layer (ETL) [88]. As a result of balanced charge injection and transport, a forward-viewing LE of 7.2–10.4 cd A^{-1} and a forward-viewing PE of 1.5–3.0 lm W^{-1} were obtained, corresponding to total LE of 15–21 cd A^{-1} and forward-viewing PE of 3.0–6.0 lm W^{-1} if all of the photons emitted out of the devices are to be collected for lighting application.

Once efficient charge injection and balanced transport could be achieved, WPLEDs based on conventional materials could exhibit high EL efficiencies. For example, Huang et al. demonstrated a highly efficient WPLEDs from incomplete transfer of energy from PFO to MEH-PPV, while the color can be readily modulated from yellow to white [with CIE coordinates of (0.36, 0.40)] by changing the concentration of MEH-PPV [21]. They found out that incorporation of a hole-blocking layer of Cs_2CO_3 at a few nanometers thick, the PE and LE of resulted WPLED were significantly enhanced up to 16 lm W^{-1} and 11.2 cd A^{-1}, respectively. The explanation for this scenario is that the thin layer of Cs_2CO_3 can only partially block the holes, and thus hole accumulation is limited. Besides, the charge-trapping effect on the MEH-PPV molecules enhances the hole accumulation more than the interface blocking effect. Thus, the combination of these effects results in better self-balanced electrons and holes in the light-emitting layer, and therefore much higher EL efficiency could be attained. This result is listed as one of the highest reported PE for fluorescent WPLEDs based on polymer blend systems.

Moreover, Zhang et al. further developed a novel method by employing a thin layer of poly[9,9-bis(6'-(diethanolamino)hexyl)-fluorene] (PFN-OH, **P55**) spin-coated from water/ethanol mixed solvent as electron-injection layer between the emitting layer and high work function metal Al cathode for WPLEDs [23]. The emitting layer consisted of two phosphors (blue-emitting [FIrpic] and an orange-emitting Os complex) doped into a PVK/OXD-7 host matrix. It was found that the device performance was strongly dependent on the morphology of the electron-injection layer and the processing solvents. Optimized devices from water/ethanol mixture showed a high LE of 20.4 cd A^{-1} and a PE of 14.5 lm W^{-1} with CIE coordinates of (0.312, 0.355), while a moderate LE of 5–6 cd A^{-1} was only achieved for the control device.

Zhang et al. also tested this PFN-OH/Al cathode in the fabrication of efficient WPLEDs based on other fluorescent host-phosphorscent dopant system. For example, by doping an ultraviolet-blue light emitting host poly[2,7-(9,9-dioctyl-fluorene)-co-1,3-(5-carbazolphenylene)] with a yellow-emitting osmium complex [Os(fptz)$_2$(dppe)] [fptz = 3-trifluoromethyl-5-(2-pyridyl)-1,2,4-triazole, dppe = cis-1,2-bis-(dipheneyl-phosphino)ethane] (**C20**) and using a blue-emitting copolymer 4-N,N-diphenylaminostilbene (PFCz-DPS1-OXD5) as the emissive layer, and PFN-OH as electron-injection layer [89], the best device performance was found to be 16.9 cd A^{-1} with CIE coordinates of (0.33, 0.34) and a maximal brightness of 22,100 cd m^{-2}.

Very recently, Huang et al. reported highly efficient all-phosphorescent WPLEDs derived from a Li_2CO_3-doped water/alcohol-soluble neutral conjugated polymer poly[9,9-bis(2-(2-(2-diethanolaminoethoxy)ethoxy)ethyl)fluorene], which was used as ETL [24]. As a result of improved electron-transporting/hole-blocking properties of the ETL, highly efficient WPLEDs with a peak LE of 36.1 cd A^{-1} and a peak PE of 23.4 lm W^{-1} were achieved from an emissive layer, in which the polymer host was doped with the sky-blue-emitting [FIrpic] (**C17**) and an orange phosphorescent osmium complex (**C14**). The reported PE is among one of the highest values reported in WPLEDs.

Conjugated oligoelectrolytes may offer some advantages over their polymeric analogs, in that their molecular structures are precisely defined and their synthesis yields no batch-to-batch variations. Xu et al. demonstrated that conjugated oligoelectrolyte can be used as good electron-injection/transport materials for efficient WPLEDs [70]. With incorporation of a thin layer of hexacationic fluorene trimer with (*N*,*N*,*N*-trimethylammonium)hexyl substituents and tetrakis-(1-imidazolyl) borate counterions (FFF-BIm4, **C21**, Scheme 10) as ETL between the emissive layer and Al cathode, it was found that the device performance can be enhanced

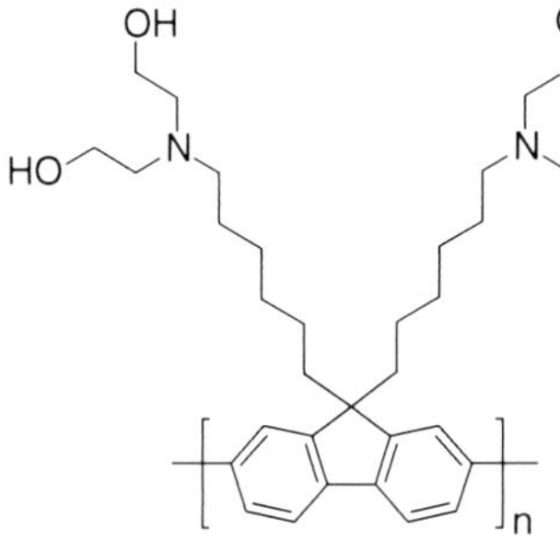

P55 PFN-OH[24]

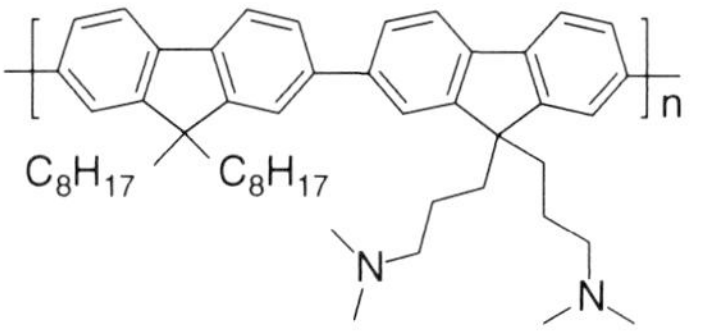

P56 PFN[71]

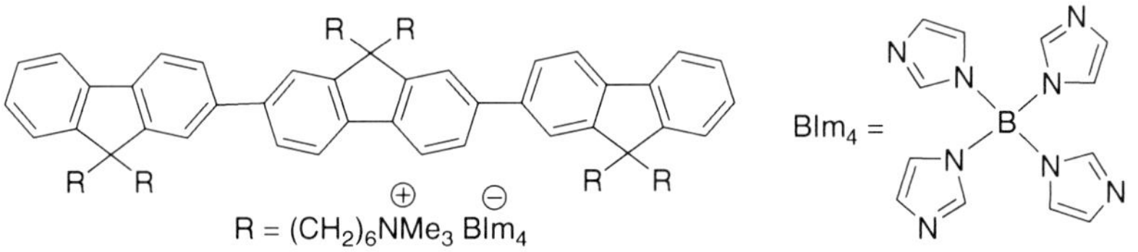

C21 FFF-BIm4[70]

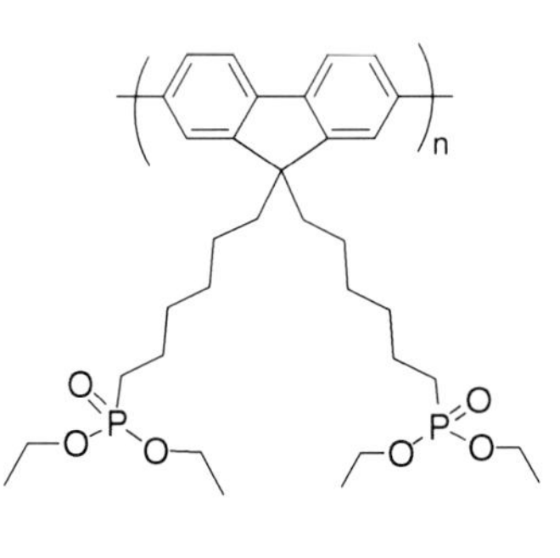

P57 PF-EP[90]

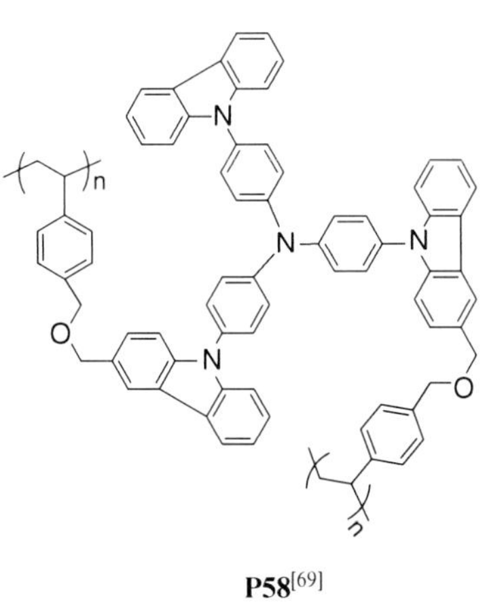

P58[69]

Scheme 10 Chemical structures of charge-injection/transport polymers or oligomers to a level comparable to that of devices based on low work function metal or those incorporated with conjugated polymer or polyelectrolyte as ETL. Efficient WPLEDs were fabricated based on a composition of [FIrpic]:[Ir(piq)] = 30:1 (by weight) in PVK/OXD-7 host matrix. It is interesting to note that the solvent mixture used for the ETL deposition can dramatically influence the device performance. Best results were obtained when a ratio between methanol and water was fixed at 3:1 for the solvent mixture. A nearly pure white color with CIE coordinates of (0.31, 0.34) was recorded with a high LE of 15.1 cd A^{-1}.

Similar to the previous works, we confirmed that highly efficient WPLEDs can be obtained from multilayer structure formed by solution-processed technique, in which an alcohol/water-soluble copolymer, poly[(9,9 bis(3′ (*N,N* dimethylamino) propyl)-2,7-fluorene)-*alt*-2,7-(9,9-dioctyl-fluorene)] (PFN, **P56**) was incorporated as ETL and Al as cathode [71]. With a device configuration of ITO/PEDOT:PSS/ PVK:OXD-7:[FIrpic] (5 wt%):[Ir(piq)] (0.25 wt%)/ETL/Al, the optimized device with a peak LE of 18.5 cd A^{-1} under forward viewing was achieved, which is comparable to that of the device with the same emissive layer but with low work function metal Ba cathode (16.6 cd A^{-1}), implying that multilayer configuration is an effective approach for producing low-cost and large-area white PLEDs by spin-coating. White emission with CIE coordinates of (0.321, 0.345) at current 10 mA cm^{-2} was observed. In addition, the emission color of the obtained devices showed minor color shift upon change of applied current density; for example, as the current density increased from 5 mA cm^{-2} to 10, 20, and 50 mA cm^{-2}, CIE coordinates shifted slightly from (0.325, 0.345) to (0.321, 0.347).

Similarly, Niu et al. demonstrated the realization of efficient multilayer WPLEDs via novel device structure, in which a neutral alcohol-soluble polyfluorene derivative was used as both blue-emitting layer and ETL in combination with another polyfluorene derivative as another emissive layer [90]. The two polymers used were poly[9,9-di-*n*-octylfluorene-*co*-4,7-bis(4-[*N*-phenyl-*N*-(4-methylphenyl) amino]phenyl)-2,1,3-benzothiadiazole] (PF-BT05) and poly[(9′9-bis(6′-diethoxyl-phosphorylhexyl)fluorene) (PF-EP, **P57**), and their chemical structures are shown in Scheme 10. Since PF-BT05 and PF-EP emissive layer were spin-coated from toluene and water, respectively, interface mixing problem was avoided. With high work function metal Al as cathode, an LE of 16.9 cd A^{-1} and a PE of 11.1 lm W^{-1} were achieved for the device; meanwhile, CIE coordinates of (0.38, 0.37) were recorded at a driving voltage of 7 V. Niu et al. also reported the realization of highly efficient WPLEDs by using a hole-injection/ transporting layer formed by the thermal crosslinking of a tris(4-carbazole)tri-phenylamine (TCTA) with two crosslinkable vinylbenzyl (VB) ether groups (**P58**) on top of a PEDOT:PSS layer [69]. White emission was realized with three phosphorescent emitters doped into PVK, which are [FIrpic] (**C17**) for blue emission, *fac*-tris(2-phenylpyridine)iridium [Ir(ppy)$_3$] for green emission, and [Os(bptz)$_2$L$_2$] (L = PPh$_2$Me) (**C20**) for red emission. Integrating the conductive hole-injection layer and crosslinkable HTL together with effective electron-blocking/exciton confinement by incorporation of tris(*N*-phenylbenzimidazol-2-yl) benzene (TPBI) at the cathode side resulted in efficient WPLEDs. A high LE of 10.9 cd A^{-1} and a peak PE of 6.2 lm W^{-1} were recorded at forward viewing with CIE coordinates changed from (0.379, 0.367) to (0.328, 0.351) at different operation voltages. Table 6 summarizes device performance of WPLEDs listed in this section.

Table 6 Summary of device performance of WPLEDs from various approaches

Device type	CIE (x, y)	λ_{max} (nm)	LE (cd A^{-1})	EQE (%)	PE (lm W^{-1})	References	Remarks
WPLEDs from fluorescent polymer blends	0.32, 0.33	426, 562	4.1	1.6	1.15	[60]	
	0.31, 0.37	438, 512, 616	7.6			[73]	
	0.34, 0.35	450, 550, 573	14.0	6.9	7.6	[74]	CRI = 72–79, CCT = 2,500–6,500 K
	0.32, 0.36		3.2			[77]	
WPLEDs from phosphorescent polymer blends	0.34, 0.36			4.7		[78]	
	0.34, 0.35		1.3			[80]	
WPLEDs using fluorescent dye-dispersed polymer blend	0.33, 0.34	430, 500, 552–566	7.5	2.5		[56]	
	0.32, 0.36		3.6		1.6	[57]	
	0.33, 0.43		17.9		16.3	[58]	
	0.35, 0.40	425, 447, 555	9.0			[59]	
	0.29, 0.38	495, 560	1.8			[62]	
WPLEDs using phosphorescent dyes-dispersed polymer blend	0.35, 0.37	420, 470, 484, 568	4.6	2.8		[63]	
	0.34, 0.35	432, 459, 542, 600	5.3		3.0	[64]	CRI = 92
	0.33, 0.32	420, 500, 590	4.3			[65]	CRI = 86–92, CCT = 4,600–6,400 K
	0.33, 0.33	423, 510, 610	9.0		5.5	[66]	
	0.33, 0.34	465, 602	13.2	6.1		[67]	
	0.34, 0.34	460, 518, 618	12.5	3.2		[68]	CRI = 83–86
	0.34, 0.47	472, 512, 620	24.3	14.4	9.5	[22]	CRI = 77, CCT = 5,010 K
	0.33, 0.36	472, 512, 620	16.1	10.0	6.3	[22]	CRI = 52, CCT = 5,896 K
WPLEDs from multiple emissive layers	0.33, 0.32	472, 620	4.4	3.0		[84]	
	0.32, 0.34	415, 477, 593	4.4			[85]	
			10.0			[86]	
WPLEDs with functional charge injection/ transport layers	0.36, 0.40		11.2	6.0	16.0	[21]	
	0.31, 0.36	472, 610	20.4	11.6	14.5	[23]	
	0.38, 0.38		36.1		23.4	[24]	
	0.38, 0.37		10.9	5.9	6.15	[69]	
	0.31, 0.34	472, 610	15.1			[70]	
	0.32, 0.35	472, 512, 620	18.5			[71]	
	0.33, 0.33	420, 500, 590	10.4		3.0	[88]	CRI = 86–92, CCT = 4,600–6,400 K
	0.33, 0.34	436, 564	16.9			[89]	
	0.38, 0.37	432, 456, 580	16.9		11.1	[90]	

5 Conclusions

The results summarized in this chapter clearly demonstrate that the performance of WPLEDs has been improved significantly in the past decade due to the progress of synthesis of novel WLEPs, optimization of device structure, and application of triplet emitters in WPLEDs. Particularly, the PE of WPLEDs can exceed that of incandescent lighting bulb, thus representing a significant advancement meet toward practical applications. However, even for the most efficient WPLEDs nowadays, their PEs are still too low for practical lighting applications. Apart from the relatively lower efficiencies, there are many challenges that remained to be addressed. Among these, as in the case of WOLEDs, the peak efficiency of WPLEDs cannot be sustained at high electrical current densities, limiting their light output per unit area which is important for lighting applications. Operation lifetimes of WPLEDs have almost not been tested and reported nowadays. It is expected that with future efforts to further enhance the device PE for WPLEDs, the studies on stability (including shelf and stress stability on device efficiency and EL spectra) will be placed on the agenda in many laboratories.

Acknowledgments The authors are grateful to the Ministry of Science and Technology Project (No. 2009CB623602) and the Natural Science Foundation of China (Project No. 50433030 and U0634003) for the financial support.

References

1. D'Andrade BW, Forrest SR (2004) Adv Mater 16:1585
2. Misra A, Kumar P, Kamalasanan MN et al (2006) Semicond Sci Technol 21:R35
3. Sun YR, Giebink NC, Kanno H et al (2006) Nature 440:908
4. Schwartz G, Pfeiffer M, Reineke S et al (2007) Adv Mater 19:3672
5. Su SJ, Gonmori E, Sasabe H et al (2008) Adv Mater 20:4189
6. UDC website (http://www.universaldisplay.com)
7. Color temperature. http://en.wikipedia.org/wiki/Color_temperature. Accessed 6 Apr 2009
8. Color rendering index. http://en.wikipedia.org/wiki/Color_rendering_index. Accessed 16 Apr 2009
9. International Commission on Illumination. http://en.wikipedia.org/wiki/International_Commission_on_Illumination. Accessed on 18 Feb 2009
10. Standard illuminant. http://en.wikipedia.org/wiki/Standard_illuminant. Accessed 10 Apr 2009
11. Cao Y, Parker ID, Yu G et al (1999) Nature 397:414
12. Shuai Z, Beljonne D, Silbey RJ, Bredas JL (2000) Phys Rev Lett 84:131
13. Greenham NC, Friend RH, Bradley DDC (1994) Adv Mater 6:491
14. Baldo MA, O'Brien DF, You Y et al (1998) Nature 395:151
15. Adachi C, Baldo MA, Thompson ME et al (2001) J Appl Phys 90:5048
16. Forrest SR, Bradley DDC, Thompson ME (2003) Adv Mater 15:1043
17. Gong X, Robson MR, Ostrowski JC et al (2002) Adv Mater 14:581
18. Ohta N, Robertson AR (2005) Colorimetry. Wiley, West Sussex, England
19. D'Andrade BW, Holmes RJ, Forrest SR (2004) Adv Mater 16:624

20. Luminous efficacy. In: Wikipedia, The Free Encyclopedia. http://en.wikipedia.org/w/index.php?title=Luminous_efficacy&oldid=257869605. Accessed 16 Dec 2008
21. Huang J, Li G, Wu E et al (2006) Adv Mater 18:114
22. Wu HB, Zou JH, Liu F et al (2008) Adv Mater 20:696
23. Zhang Y, Huang F, Chi Y et al (2008) Adv Mater 20:1565
24. Huang F, Shih PI, Shu CF et al (2009) Adv Mater 21:361
25. Tu GL, Zhou QG, Cheng YX et al (2004) Appl Phys Lett 85:2172
26. Sun ML, Niu QL, Du B et al (2007) Macromol Chem Phys 208:988
27. Lee SK, Jung BJ, Ahn T et al (2007) J Polym Sci A Polym Chem 45:3380
28. Hsieh BY, Chen Y (2008) J Polym Sci A Polym Chem 46:3703
29. Liu J, Gao BX, Cheng YX et al (2008) Macromolecules 41:1162
30. Chien CH, Shin PI, Shu CF (2007) J Polym Sci A Polym Chem 45:2938
31. Park MJ, Lee J, Jung IH et al (2008) Macromolecules 41:9643
32. Tu GL, Mei CY, Zhou QG et al (2006) Adv Funct Mater 16:101
33. Liu J, Zhou QG, Cheng YX et al (2006) Adv Funct Mater 16:957
34. Liu J, Guo X, Bu LJ et al (2007) Adv Funct Mater 17:1917
35. Liu J, Shao SY, Chen L et al (2007) Adv Mater 19:1859
36. Liu J, Cheng YX, Xie ZY et al (2008) Adv Mater 20:1357
37. Xu YH, Guan R, Jiang JX et al (2008) J Polym Sci A Polym Chem 46:453
38. Mei CY, Ding JQ, Yao B et al (2007) J Polym Sci A Polym Chem 45:1746
39. Furuta PT, Deng L, Garon S et al (2004) J Am Chem Soc 126:15388
40. Deng L, Furuta PT, Garon S et al (2006) Chem Mater 18:386
41. Lee SK, Hwang DH, Jung BJ et al (2005) Adv Funct Mater 15:1647
42. Liu J, Zhou QG, Cheng YX et al (2005) Adv Mater 17:2974
43. Luo J, Li X, Hou Q et al (2007) Adv Mater 19:1113
44. Chuang CY, Shih PI, Chien CH et al (2007) Macromolecules 40:247
45. Wang F, Wang L, Chen JW et al (2007) Macromol Rapid Commun 28:2012
46. Liu J, Xie ZY, Cheng YX et al (2007) Adv Mater 19:531
47. Liu J, Chen L, Shao SY et al (2007) Adv Mater 19:4224
48. Lee PI, Hsu SLC, Lee RF (2007) Polymer 48:110
49. Zhen HY, Xu W, Yang W et al (2006) Macromol Rapid Commun 27:2095
50. Chen QL, Liu NL, Ying L et al (2009) Polymer 50:1430
51. Lee PI, Hsu SLC, Lee JF (2008) J Polym Sci A Polym Chem 46:464
52. Zhang K, Chen Z, Yang CL et al (2008) J Mater Chem 18:291
53. Jiang JX, Xu YH, Yang W et al (2006) Adv Mater 18:1769
54. Wu FI, Yang XH, Neher D et al (2007) Adv Funct Mater 17:1085
55. Chien CH, Liao SF, Wu CH et al (2008) Adv Funct Mater 18:1430
56. Fan SQ, Sun ML, Wang J et al (2007) Appl Phys Lett 91:213502
57. Xu QF, Duong HM, Wudl F et al (2004) Appl Phys Lett 85:3357
58. Huang JS, Hou WJ, Li JH et al (2006) Appl Phys Lett 89:133509
59. Peng KY, Huang CW, Liu CY et al (2007) Appl Phys Lett 91:093502
60. Shih PI, Tseng YH, Wu FI et al (2006) Adv Funct Mater 16:1582
61. Kim JH, Herguth P, Kang MS et al (2004) Appl Phys Lett 85:1116
62. Liu F, Tang C, Chen QQ et al (2009) J Phys Chem C 113:4641
63. Attar HAA, Monkman AP, Tavasli M et al (2005) Appl Phys Lett 86:121101
64. Niu XD, Ma L, Yao B et al (2006) Appl Phys Lett 89:213508
65. Gong X, Ma W, Ostrowski JC et al (2004) Adv Mater 16:615
66. Xu YH, Peng JB, Jiang JX et al (2005) Appl Phys Lett 87:193502
67. Shih PI, Shu CF, Tung YL et al (2006) Appl Phys Lett 88:251110
68. Kim TH, Lee HK, Park O et al (2006) Adv Funct Mater 16:611
69. Niu YH, Liu MS, Ka JW et al (2007) Adv Mater 19:300
70. Xu YH, Yang RQ, Peng JB et al (2009) Adv Mater 21:584
71. An D, Zou JH, Wu HB et al (2009) Org Electron 10:299

72. Hsieh BY, Chen Y (2009) J Polym Sci A Polym Chem 47:833
73. Niu QL, Xu YH, Jiang JX et al (2007) J Lumin 126:531
74. Zou JH, Liu J, Wu HB et al (2010) Org Electron 10:843
75. Lee BK, Ahn T, Cho NS et al (2007) J Polym Sci A Polym Chem 45:1199
76. Tan ZA, Tang RP, Sun QJ et al (2007) Thin Solid Films 516:47
77. Sun QJ, Hou JH, Yang CH et al (2006) Appl Phys Lett 89:153501
78. Tokito S, Suzuki M, Sato F et al (2003) Org Electron 4:105
79. Kappaun S, Eder S, Sax S et al (2006) J Mater Chem 16:4389
80. Xiong Y, Zhang Y, Zhou JL et al (2007) Chin Phys Lett 24:3547
81. Chao CI, Chen SA (1998) Appl Phys Lett 73:426
82. Sun QJ, Fan BH, Tan ZA et al (2006) Appl Phys Lett 88:163510
83. Lee YZ, Chen XW, Chen MC et al (2001) Appl Phys Lett 79:308
84. Xu YH, Peng JB, Mo YQ et al (2005) Appl Phys Lett 86:163502
85. Zhou Y, Sun QJ, Tan ZA et al (2007) J Phys Chem C 111:6862
86. Sun QJ, Chang DW, Dai LM et al (2008) Appl Phys Lett 92:251108
87. Salaneck WR, Seki K, Kahn A (2001) In: Pireaux JJ (ed) Conjugated polymer and molecular interfaces: science and technology for photonic and optoelectronic applications. Dekker, New York
88. Gong X, Wang S, Moses D et al (2005) Adv Mater 17:2053
89. Zhang Y, Huang F, Jen AKY et al (2008) Appl Phys Lett 92:063303
90. Niu XD, Qin CJ, Zhang BH et al (2007) Appl Phys Lett 90:203513

Phosphorescent Platinum(II) Complexes for White Organic Light-Emitting Diode Applications

Chi-Chung Kwok, Steven Chi Fai Kui, Siu-Wai Lai, and Chi-Ming Che

Contents

1 Introduction 80
2 Working Principle for WOLEDs 81
   2.1 Mixing Red–Green–Blue 81
   2.2 Mixing Complementary Colors 82
   2.3 Emitting Materials with Complementary Colors 83
3 Phosphorescent Platinum(II)-Based Materials 83
   3.1 Bidentate Platinum(II) Complexes 85
   3.2 Tridentate Platinum(II) Complexes 90
   3.3 Tetradentate Platinum(II) Complexes 95
   3.4 Binuclear Platinum(II) Complexes 97
   3.5 Polymeric Materials 100
4 Summary 103
References 103

Abstract The applications of phosphorescent platinum(II) complexes in white organic light-emitting diode (WOLED) are discussed. White electroluminescence formed by complementary colors mixing has been achieved by employing phosphorescent platinum(II) complexes as dopants. The approach is to mix triplet monomer emissions of the platinum(II) dopant complexes at orange-red region with a blue-emitting component or, alternatively, to mix the emissions from both monomer and aggregate states of the same platinum(II) complex in the blue-green (λ_{max} ~ 480nm) and orange-red (λ_{max} ~ 600 nm) regions. Platinum(II) material-based WOLEDs could be fabricated from both thermal deposition and solution process, since polymeric WOLED materials could be prepared by incorporating platinum(II) complexes in polymer backbone. The WOLEDs fabricated from platinum(II) complexes exhibit good Commission Internationale de l'Eclairage, color-rendering index, and device efficiency, which may find potential applications for solid-state lighting.

C.-C. Kwok, S.C.F. Kui, S.-W. Lai, and C.-M. Che (✉)
Department of Chemistry, The University of Hong Kong, Pokfulam Road, Hong Kong S. A. R., China
e-mail: cmche@hku.hk

1 Introduction

Since Tang and Van Slyke reported the first organic light-emitting diode (OLED) in 1987 [1], the advantages of wide viewing angle, fast response time, high contrast ratio, and wide temperature operation range in OLEDs have rendered it topical in the research both in the academic and industrial communities. Intensive efforts have been devoted to the development of OLEDs for the new generation display panels. Ranging from monochromatic passive matrix (PM) display screens in MP3 players, small-sized (< 3.5 inch) subdisplays in mobile phones, and active matrix OLED (AMOLED) in mobile phones and digital cameras, OLEDs are widely used in pocket-sized consumer electronics [2, 3]. Sony Corporation launched an 11 inch OLED TV (XEL-1) in 2007, and a 15-in. OLED TV (15EL9500) has been provided by LG in 2010. Both the televisions have ultra-thin panel of 3 mm and 1.7 mm, respectively, and high contrast ratio of 1,000,000:1, revealing that the technologies for AMOLED panel become mature for production [4, 5].

The latest global trends in energy saving have promoted vibrant research activities at the research institutes, universities, and industries toward the development of white organic light-emitting diode (WOLED) with low power consumption for the applications in solid-state and flexible lighting. In 1994, Kido et al. demonstrated a dip-coated WOLED using three fluorescent dyes, namely 1,1,4,4-tetraphenyl-1,3-butadiene (TPB), coumarin 6, and 4-(dicyanomethylene)-2-methyl-6-(p-dimethylaminostyryl)-4H-pyran (DCM), which were doped in a poly(N-vinylcarbazole) (PVK) layer [6]. The electroluminescence (EL) spectrum covers a wide range of the emission colors in the visible region with peak wavelengths at 450, 510, and 550 nm, which correspond to the emission from TPB, coumarin 6, and DCM, respectively. Later, Kido et al. fabricated a WOLED through thermal deposition using N,N'-diphenyl-N,N'-bis(3-methylphenyl)-(1,1'-biphenyl)-4,4'-diamine (TPD), tris(8-hydroxyquinolinato)aluminum (Alq_3), and Nile Red as emitters [7]. The EL spectrum shows a broad coverage of the visible region with three peaks at wavelengths of 410, 520, and 600 nm, and the device produced a luminous efficiency of 0.5 lm W^{-1}. Over the past decade, both the efficiency and lifetime of WOLEDs have been improved tremendously. The prototype device with luminous efficiency of 60 lm W^{-1} and lifetime of > 60,000 h has been demonstrated by Kodak, revealing that WOLEDs are good alternatives to incandescent or fluorescent lighting. Furthermore, they offer the possibility of changing into various shapes and sizes for customized applications.

One of the keys to improve the efficiencies of OLEDs is to make use of phosphorescent-emitting materials as the excitons generated by the recombination of holes and electrons have a statistical singlet:triplet ratio of 1:3. Through intersystem crossing from the singlet to the triplet excited state, phosphorescent-emitting materials could utilize both singlet and triplet excitons for light-emitting applications. This chapter summarizes the works done on using phosphorescent platinum (II) complexes for WOLED applications and focuses on the performance characteristics of these devices.

2 Working Principle for WOLEDs

2.1 Mixing Red–Green–Blue

White light can be generated by mixing the three primary colors (red, green, and blue; RGB). The electroluminescence from devices of mixing RGB dyes covers almost the entire visible spectral region, and the white emission can be used for lighting and backlight applications for LCD and other color filter devices. This approach was first realized by Kido et al. in 1995 [7], and the WOLED fabricated by them displayed low turn-on voltage and high luminance. In 2008, Kim et al. used phosphorescent iridium(III) complexes as emitting materials for RGB mixing in the WOLED fabrication; the three dopants are tris[1-phenylisoquinolinato-C^2,N]iridium(III) [Ir(piq)$_3$], tris-fac-[2-(3-methylcyclohex-1-enyl)pyridine]iridium(III) [Ir(mchpy)$_3$], and bis[(4,6-difluorophenyl)-pyridinato-N,$C^{2'}$]picolinatoiridium(III) [FIrpic] as red-, green-, and blue-emitting materials, respectively. Figure 1 depicts the EL spectra of the WOLED at different luminances of 10, 100, 1,000, and 5,000 cd m^{-2}. The WOLED exhibited the maximum device efficiency of 11.7%, and the Commission Internationale de l'Eclairage (CIE) coordinates of (0.38, 0.43) [8]. Moreover, the WOLED showed high color stability with increasing operating current and voltage.

In general, it is not an easy task to maintain the color purity and stability of WOLEDs fabricated by mixing RGB dopants. The performances of devices usually change significantly with small variation in dopant concentration, thickness of the emissive layers [9, 10], operating voltage [11], and current [12]. Furthermore, different electroluminescent lifetimes of the various dopants would alter the color purity of the WOLEDs over time.

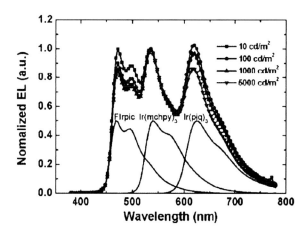

Fig. 1 EL spectra of RGB-mixing WOLED using iridium(III) complexes as emitting materials. Reproduced with permission from [8]. Copyright 2008, WILEY-VCH Verlag GmbH & Co. KGaA, Weinheim

2.2 Mixing Complementary Colors

Apart from the approach of mixing RGB colors, white light can be generated by mixing two complementary colors as depicted in the CIE 1931 chromaticity diagram (Fig. 2). Visible color can be represented by two-dimensional coordinates (x, y) in the CIE chromaticity diagram. If a straight line is drawn through the white region roughly from (0.25, 0.25) to (0.35, 0.35), two colors at the ends of the line with appropriate relative intensity can be combined to give white light. By using this method, only two emitting materials are needed, and one of the emitting materials can be the host material of a doping device. In this case, the color purity can be more easily controlled than those in the RGB-mixing devices. Although the EL spectrum of white light produced by this method may not fully cover the whole visible region, it is already good enough for lighting applications. From the time of the device reported by Forrest et al. [13], most of the WOLED devices have been fabricated by this method [14–17]. Recently, Kido et al. reported a two-color-mixing WOLED with power efficiency of 55 lm W^{-1} using [FIrpic] and iridium (III) bis-(2-phenylquinoly-$N,C^{2'}$)dipivaloylmethane [PQ2Ir] as blue- and orange-emitting materials, respectively [18].

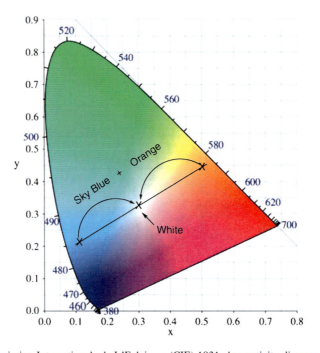

Fig. 2 Commission Internationale de L'Eclairage (CIE) 1931 chromaticity diagram

2.3 Emitting Materials with Complementary Colors

The most popular combination of complementary colors in WOLED applications is by mixing blue to sky-blue with orange to red electroluminescence. For blue-light emitter, fluorescent materials such as TPD, N,N'-bis(1-naphthyl)-N,N'-diphenyl-1,1'-biphenyl-4,4'-diamine (NPB), N-(4-(((1E,22E)-4-((E)-4-((2-ethyl-6-tolyl)(phenyl)amino)styryl)styryl)phenyl)-N-(2-ethyl-6-tolyl)benzenamine (BUBDU-1), 4-{4-[N-(1-naphthyl)-N-phenylaminophenyl]}-1,7-diphenyl-3,5-dimethyl-1,7-dihydro-dipyrazolo-[3,4-b;4'3'-e]pyridine (PAP-NPA), 1,4-bis(1,1-diphenyl-2-ethyenyl) benzene (PEB), 4,4'-bis(2,2'-diphenylvinyl)-1,1'-biphenyl (DPVBi), and dinaphthylanthracene (DNA) are commonly used (the chemical structures are shown in Scheme 1). Phosphorescent materials have also been used as blue-emitting materials in WOLEDs; for instance, high-performance WOLEDs described by Kido [18] were fabricated using [FIrPic] as blue emitter. Single emissive layer WOLED can be fabricated by doping suitable concentration of orange light-emitting material into these blue-emitting materials [19–24].

For the yellow to red components in WOLEDs, fluorescent materials were often used such as DCM, 5,6,11,12-tetraphenylnaphthacene (rubrene), 4-(dicyanomethylene)-2-$tert$-butyl-6-(1,1,7,7-tetramethyljulolidin-4-yl-vinyl)-4H-pyran (DCJTB), and their derivatives. High-performance WOLED with efficiency of 17.1 cd A^{-1} has been realized by Lin et al. using fluorescent orange-emitting material tetra-($tert$-butyl)-rubrene (TBRb) [16].

However, the luminous efficiency of devices comprising fluorescent emitters is restricted by the yield of singlet excitons. More recently, the development of phosphorescent materials as dopants in OLEDs has become the focus of research due to the possibility of 100% internal quantum efficiency. WOLEDs based on phosphorescent dopant in fluorescent host were reported to exhibit markedly improved device performance; device efficiency of up to 55 lm W^{-1} was described using [PQ2Ir] as orange emitter [18].

3 Phosphorescent Platinum(II)-Based Materials

Studies on the photoluminescence of platinum(II) complexes prior to 1986 had led to limited applications in OLEDs as most of these compounds are only emissive in rigid media at low temperature [25–27]. In the late 1980s, reports on platinum(II) complexes that exhibit photoluminescence in fluid solutions at ambient temperatures have been documented in the literature [28–30]. In 1987, von Zelewsky described platinum(II) diimine complexes with excited state having metal-to-ligand charge transfer (^{3}MLCT) character [30], and since then the photoluminescent properties of platinum(II) complexes have received considerable attention. We and Eisenberg reported the luminescent platinum(II) acetylide complexes bearing aromatic diimine ligands [31, 32], which were eventually used as emitting materials

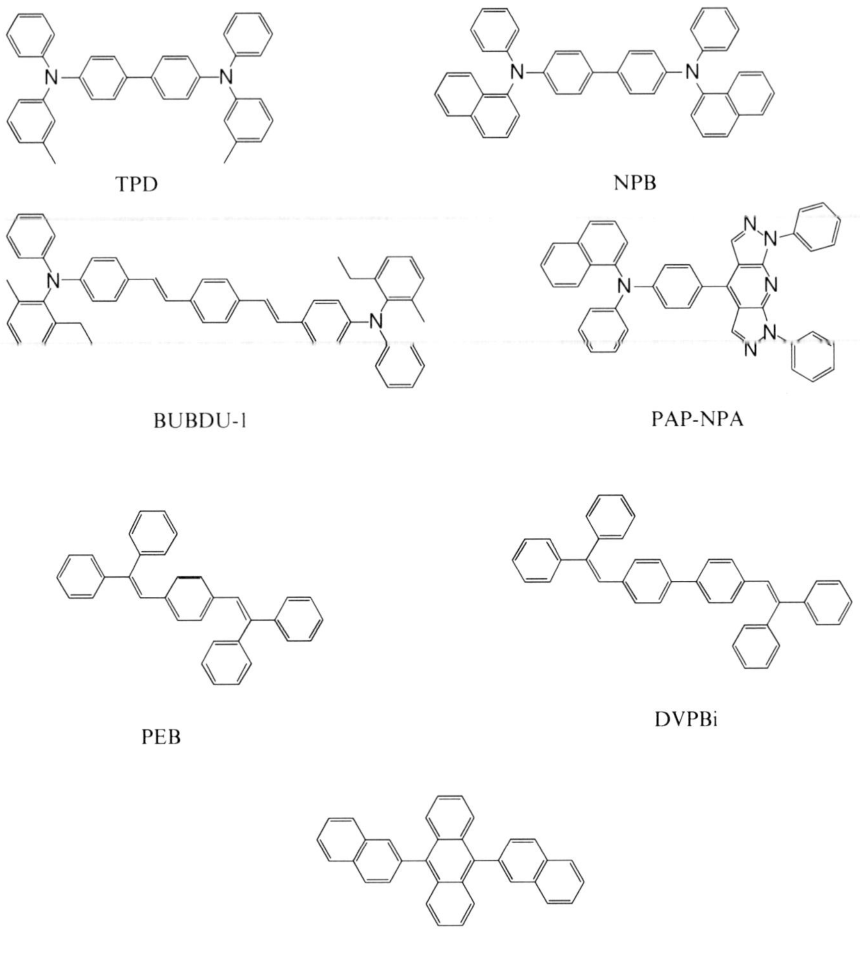

Scheme 1 Chemical structures of commonly used fluorescence blue-emitting materials: TPD, NPB, BUBDU-1, PAP-NPA, PEB, DVPBi, and DNA

in OLEDs [32]. Subsequently, phosphorescent platinum(II) complexes have been extensively studied and widely used in OLED applications. High-performance OLED with 51.8 cd A^{-1} was reported by using platinum(II) 3-(6'-(2''-naphthyl)-2'-pyridyl)isoquinolinyl [(RC^N^N)PtCl] complexes as emitting materials [33]. Apart from the monomeric MLCT emission, the low-energy excited states of phosphorescent platinum(II) complexes include intraligand (IL) states of "excimer" type [$\sigma^*(\pi) \rightarrow \sigma(\pi^*)$] and oligomeric [$d\sigma^*(dz^2) \rightarrow \sigma(\pi^*)$] (MMLCT) in solution of high complex concentration or in solid state. Using MMLCT emissions as the operating principle, OLEDs with high doping concentration of platinum(II) complexes in the emissive layer were fabricated, and were found to show red to near-infrared emission with high device efficiencies [34, 35]. As mentioned in Sect. 2.3, WOLEDs can be achieved by a combination of blue fluorescent host with red-emitting material. Since most of the phosphorescent platinum(II) complexes show yellow to red emissions, they are good candidates as phosphorescent dopants in WOLED application. More detailed descriptions of using phosphorescent platinum(II) complexes in WOLEDs will be discussed in the following sections.

3.1 Bidentate Platinum(II) Complexes

3.1.1 Cyclometalated Platinum(II) β-diketonato [(C^N)Pt(O^O)] Complexes

In 2002, Thompson et al. reported a series of [(C^N)Pt(O^O)] complexes (C^N = monoanionic 2-phenylpyridyl, ppy; 2-(2′-thienylpyridyl), thpy; etc.) that show low-energy absorption at 350–450 nm with ε values of 2,000–6,000 $dm^3\ mol^{-1}cm^{-1}$, attributable to MLCT transition [36]. As the HOMOs of [(C^N)Pt(O^O)] complexes consist of a mixture of phenyl, platinum and O^O characters, the emission λ_{max} could be tuned hypsochromically by incorporating fluorine atoms at the 4′ and 6′ positions of the phenyl ring (i.e., F_2C^N = 2-(4′,6′-difluorophenyl)pyridinato-N, $C^{2'}$). The [(F_2C^N)Pt(O^O)] complexes showed blue emission with λ_{max} at ~466 nm in 2-methyltetrahydrofuran solution, which was originated from mixed ^{3}LC (ligand-centered) and MLCT excited states [36]. The emission characteristics of these complexes can be governed by the nature of the cyclometalated ligand.

To achieve excimer emission from these cyclometalated platinum(II) complexes, the steric bulkiness of β-diketonato ligand is important in controlling the close proximity of the planar molecules. For example, the molecular packing diagram of [(thpy)Pt(2,2,6,6-tetramethyl-3,5-heptanedionato-O,O)] bearing bulky diketonato ligand shows long intermolecular $Pt^{II}-Pt^{II}$ separation of 4.92 Å in solid state, revealing no metal–metal interactions [36]. Brooks et al. reported using [(F_2C^N)Pt(acac)] complex with less bulky acac ligand to afford the low-energy excimer emission with λ_{max} 570 nm in 4,4′-N,N′-dicarbazole-diphenyl (CBP) doping thin films [37, 38]. The low-energy emission may also be originated from $^3[\pi^*, d\sigma^*]$ (MMLCT) excited state. White light electrophosphorescence was obtained by combining orange excimer emission in [(F_2C^N)Pt(acac)] complexes (Scheme 2) with blue monomer emission from iridium(III) complex [FIrpic]; the device structure of WOLED was ITO/poly(3,4-ethylene-dioxythiophene):poly(styrene sulfonate) (PEDOT:PSS)/NPB (30 nm)/CBP:FIrpic/**1** (30 nm)/BCP (50 nm)/LiF/Al (Device 1, Fig. 3).

Device 1 shows the maximum external quantum efficiency and color-rendering index (CRI) of 4% and 78, respectively. By controlling the doping concentration of **2** in CBP, white emission with CIE of (0.33, 0.31) was observed and the device structure of ITO/NPB/CBP:**2**/BCP/Alq$_3$/LiF/Al (Device 2) was fabricated [38].

However, the EL of Device 2 showed a significant contribution from the NPB emission upon an increase in the operating voltage, which in turn reduced the device efficiency due to the excitons leakage to the hole transporting layer. To confine the emission zone, an electron-blocking layer of *fac*-tris(1-phenylpyrazolato-$N,C^{2'}$)iridium(III) [Irppz] was inserted between the NPB and the emissive layer, the device with a structure of ITO/NPB (40 nm)/Irppz (20 nm)/mCP:**1** (16%, 30 nm)/BCP (15 nm)/Alq$_3$ (20 nm)/LiF (0.1 nm)/Al (100 nm) (Device 3) was fabricated [38]. The EL spectrum of Device 3 showed sky blue monomer emission and yellow excimer emission of **1**; no emission corresponding to NPB was observed. White electroluminescence was obtained from Device 3 with maximum quantum efficiency of 6.4%. Similar results were achieved with devices using **2** as emitting material; the ratio of monomer/excimer emissions in the EL spectra did not change with applied voltage and no NPB emission was observed.

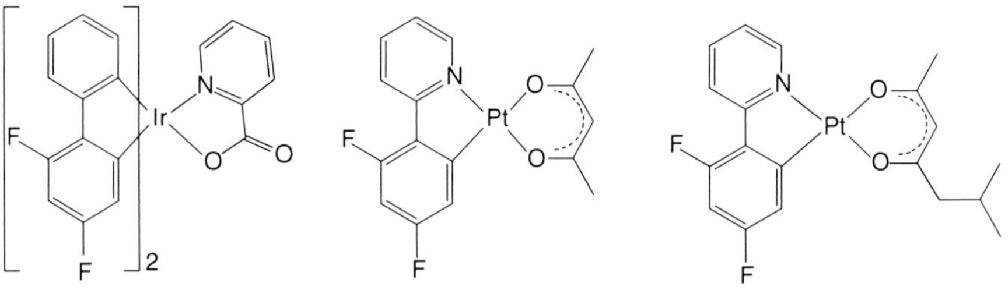

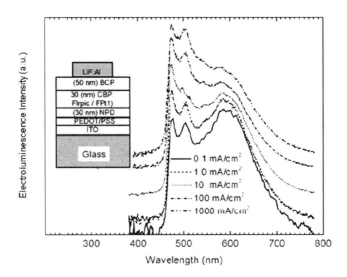

FIrpic   **[(F$_2$C^N)Pt(acac)], 1**   **2**

Scheme 2 Chemical structures of [FIrpic], [(F$_2$C^N)Pt(acac)] **1** and **2**

Fig. 3 Device structure and EL spectra of Device 1. Reproduced with permission from [37]. Copyright 2002, WILEY-VCH Verlag GmbH & Co. KGaA, Weinheim

In 2007, Jabbour et al. demonstrated efforts in modifying the device structure, thus improving the device performance and efficiency of WOLEDs using [(F_2C^N)Pt(acac)] as a phosphor excimer-based emitter. The effects on device performance by using different host materials, hole injection materials, and electron-blocking materials were analyzed [39]. Device with a structure of ITO/PEDOT:PSS/poly(N-vinylcarbazole) (PVK)/2,6-bis(N-carbazolyl)pyridine (26mCPy):**1** (12%)/BCP/LiF/Al showed the best device performance with external quantum efficiency (η_{ext}), power efficiency (η_p), luminous efficiency (η_L), CIE and CRI of 15.9%, 12.6 lm W^{-1}, 37.8 cd A^{-1}, (0.46, 0.47), 69, respectively, at 500 cd m^{-2}. Nearly 100% internal quantum efficiency was achieved at 1 cd m^{-2} (η_{ext}: 18%, η_p: 29 lm W^{-1}, and η_L: 42.5 cd A^{-1}).

In 2006, Wong et al. prepared a series of multifunctional [(C^N)Pt(O^O)] complexes bearing electron transporting, hole transporting, and emitting chromophores in one molecule (Scheme 3) [40]. As spin–orbit interaction in these complexes is reduced upon the introduction of triarylamino and oxadiazole functional groups into the phenylpyridine ligands, the solution emission spectra of **3–6** in dichloromethane solutions show dual emission peaks (Fig. 4). The high-energy emission peaks located at 429–496 nm are fluorescence in nature with lifetime of 2.2–3.1 ns, attributable to LC character in the emissive excited states. The low-energy emission peaks at ~533 nm are phosphorescence with lifetimes of 3.1–7.0 μs, which are assigned to ^{3}LC (π–π*) excited states.

By making use of both fluorescence and phosphorescence, WOLEDs with a simple device structure of ITO/NPB (60 nm)/CBP:**5** (10%, Device 4), (12%, Device 5)/2,2′,2″-(1,3,5-phenylene)-tris(1-phenyl-1H-benzimidazole) (TPBI), 30 nm)/LiF (1 nm)/Al (120 nm) were fabricated. The turn-on voltage of Devices 4 and 5 was at ~4.2 V, and the electrophosphorescence peak maximum appeared at 460 and 540 nm. Device 4 showed maximum η_{ext}, η_L, and η_p of 2.63%, 6.79 cd A^{-1}, and 4.07 lm W^{-1}, respectively, whereas the CIE coordinates of Devices 4 and 5 are (0.33, 0.39) and (0.30, 0.36), respectively (Fig. 5).

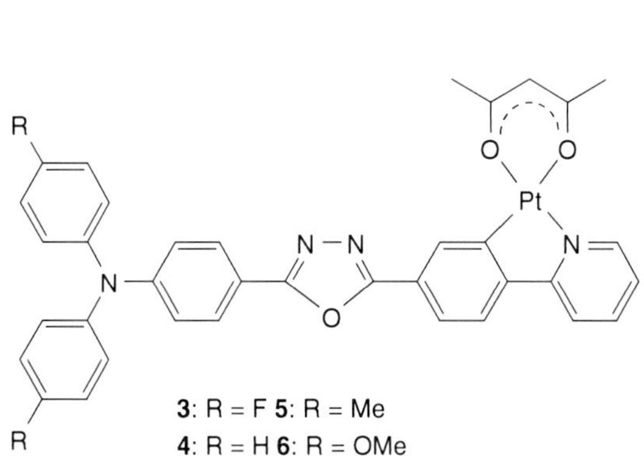

Scheme 3 Chemical structures of complexes **3–6**

3: R = F 5: R = Me
4: R = H 6: R = OMe

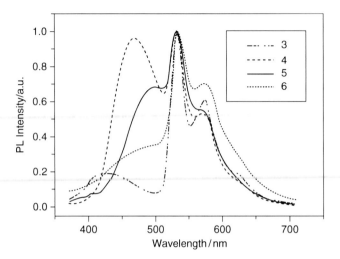

Fig. 4 Emission spectra of **3–6** in dichloromethane solution at 298 K. Reproduced with permission from [40]. Copyright 2006, American Chemical Society

3.1.2 Platinum(II) Di(pyridyl)triazolate [Pt(N^N)$_2$] Complexes

In 2009, Omary et al. presented a noncyclometalated platinum(II) complex, namely bis[3,5-bis(2-pyridyl)-1,2,4-triazolato]platinum(II) (**7** – its chemical structure is shown in Scheme 4), which shows monomer emission at λ_{max} ~ 476 nm and excimer emission at λ_{max} ~ 570 nm in OLED device. Upon mixing the excimer emission of **7** with the deep-blue emission from 4,4′-bis(9-ethyl-3-carbazovinylene)-1,1′-biphenyl (BCzVBi), white light was attained in Device 6 with the structure of ITO/NPB (40 nm)/CBP:BCzVBi (10 nm, 5%)/CBP (4 nm)/CBP:**7** (20 nm, 30%)/CBP (16 nm)/TPBI (30 nm)/Mg:Ag (100 nm). Device 6 showed power efficiency, luminous efficiency, CRI, and external quantum efficiency of 5.7 lm W^{-1}, 10.6 cd A^{-1}, 76, and 4.8%, respectively, at low brightness (50 cd m^{-2}). These efficiencies and index were almost unchanged even up to 1,100 cd m^{-2} [41, 42]. Device performances for Device 6 are summarized in Table 1.

3.1.3 Platinum(II) Bis(8-hydroxyquinolato) [PtQ$_2$] Complexes

Bis(8-hydroxyquinolinato)platinum(II) [PtQ$_2$] complexes were first prepared by Ballardini in 1978 [43], and the photophysical properties were first investigated by Traverso [44]. Subsequently, the assignment of the absorption and emission spectra and the electronic transitions were proposed to be ligand-centered $^3[l \rightarrow \pi^*]$ (l = lone pair/phenoxide) charge transfer in nature [45]. Using Shpol's-kii spectroscopy, this assignment was further confirmed by Yersin [34, 35]. Nevertheless, no practical application on [PtQ$_2$] complexes had been reported prior to our report on [PtQ$_2$]-based OLEDs in 2008 [46]. Deep red color OLEDs have been

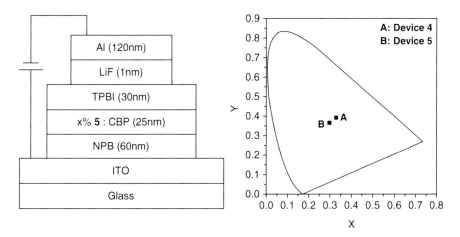

Fig. 5 Device structure (x% **5**) and CIE coordinates of Devices 4 and 5. Reproduced with permission from [40]. Copyright 2006, American Chemical Society

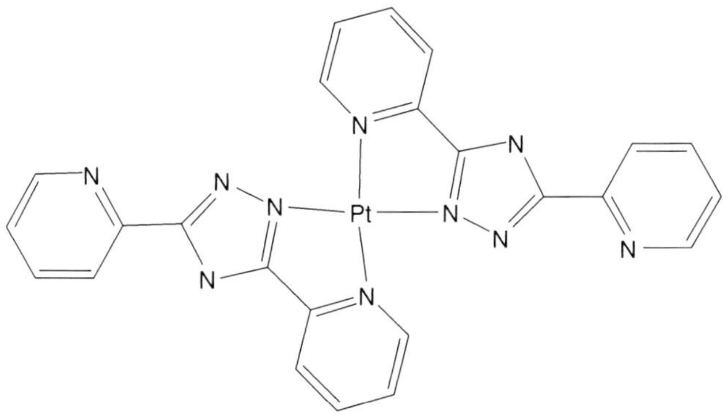

Scheme 4 Chemical structure of complex **7**

Table 1 Device performance of Device 6. Reproduced with permission from [41, 42] (Copyright 2009, American Institute of Physics)

Brightness (cd m^{-2})	Bias (V)	J (mA cm^{-2})	P.E. (lm W^{-1})	L.E. (cd A^{-1})	CRI	CIE	EQE
50 ± 2	5.8	0.44 ± 0.02	5.7 ± 0.1	10.6 ± 0.2	76	(0.28, 0.32)	4.8
100 ± 5	6.2	1.00 ± 0.05	6.0 ± 0.1	11.8 ± 0.2	76	(0.29, 0.34)	5.2
500 ± 15	7.0	3.90 ± 0.15	5.8 ± 0.1	12.9 ± 0.2	73	(0.30, 0.36)	5.4
1,100 ± 25	7.6	8.60 ± 0.25	5.4 ± 0.1	13.0 ± 0.2	72	(0.30, 0.370)	5.3

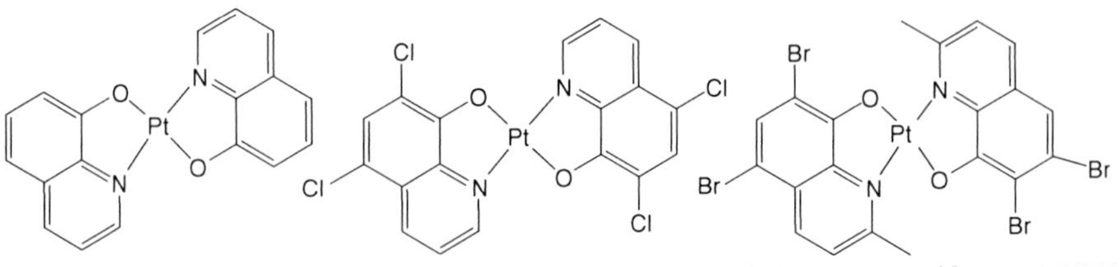

Scheme 5 Chemical structures of complexes **8–10**

fabricated from [PtQ$_2$] complexes with a maximum external quantum efficiency of 1.7% and CIE of (0.70, 0.28), revealing that [PtQ$_2$] complexes could be used as red-emitting component for WOLEDs fabrication.

Three Devices 7a–7c with the structure of ITO/NPB/bis(2-(2′-hydroxyphenyl) pyridine)beryllium (Bepp$_2$):[PtQ$_2$] (**8–10** – chemical structures are shown in Scheme 5)/LiF/Al have been fabricated. Figure 6 depicts the EL spectra of Device 7c in different operating voltages; two emission peak maxima were observed where the high-energy emission with λ_{max} at 448 nm is originated from the host material (Bepp$_2$), and the low-energy emission band with λ_{max} at 660 nm is due to the emission of **10**. Devices 7a–7c showed good color stability, and the CIE coordinates of Device 7c were found to be around (0.31, 0.22) at the operating voltages ranging from 4.2 to 11 V.

3.2 Tridentate Platinum(II) Complexes

3.2.1 Platinum(II) Di(2-pyridiyl)benzene [Pt(N^C^N)X] Complexes

Synthesis and luminescent properties of platinum(II) di(2-pyridiyl)benzene [Pt(N^C^N)X] (X = halogen, phenoxy) complexes were reported by Williams et al. in 2003 [47]. The chemical structures of the [Pt(N^C^N)X] complexes are depicted in Scheme 6. This class of complexes showed good thermal stability and high-emission quantum efficiencies ($\Phi = 0.58$ for **11**). Their emission spectra recorded in dilute dichloromethane usually display highly structured emission bands at $\lambda_{max} \sim 490$ and 525 nm, which are assigned to come from $^3\pi-\pi^*$ excited state, whereas excimer emissions were observed at $\lambda_{max} \sim 700$ nm in solution at high concentrations (1.0×10^{-4} mol dm^{-3}). With low dopant concentration in OLED fabrication, [Pt(N^C^N)X] complexes exhibited monomer emission at $\lambda_{max} \sim 500$ nm, from which high-performance green OLED with η_L of 67.1 cd A^{-1} at 5 A m^{-2} resulted in the device using **13** as emitting material [48]. The thermal deposited thin films with [Pt(N^C^N)X] complexes were found to be conductive and exhibited very high quantum efficiency red/near-infrared electroluminescence in OLEDs fabricated using the excimer emissions of these complexes [49]. As high-performance OLEDs can be fabricated from either the monomer or excimer emissions of

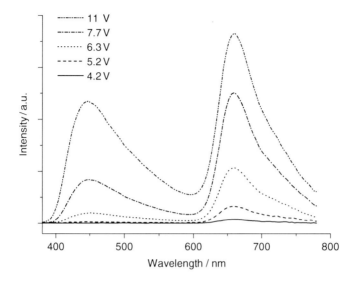

Fig. 6 EL spectra of Device 7c in various operating voltages

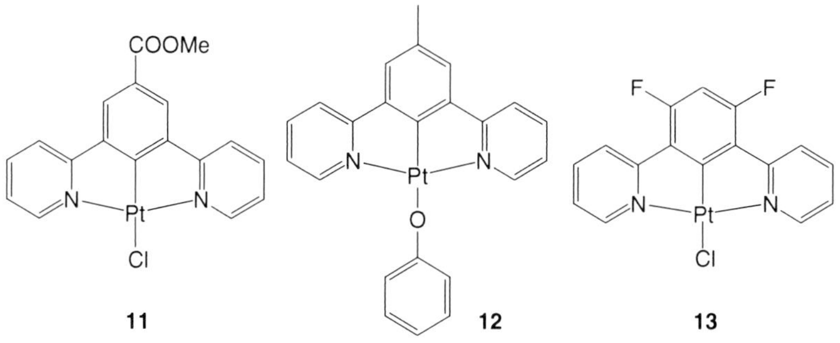

Scheme 6 Chemical structures of complexes **11–13**

[Pt(N^C^N)X] complexes, their WOLED applications were anticipated by using both the monomer and excimer emission excited states.

In 2007, Cocchi et al. described the WOLEDs by mixing monomer and excimer emissions of [Pt(N^C^N)X] complexes in a single emissive layer through the control of the doping concentration of [Pt(N^C^N)X] complexes [50]. Devices 8a–8b were fabricated with device structure of ITO/TPD:PC/CBP/CBP:OXA:**11**/OXA/Ca (Device 8a: 15% of **11** and Device 8b: 18% of **11**). Figure 7 depicts the PL spectra of monomer and excimer emission for **11** and EL spectra of Devices 8a and 8b. Both Devices 8a and 8b showed monomer and excimer emission of **11**, and the excimer is formed after the exciton is localized on the molecule via energy transfer from CBP to **11** monomer and/or hole–electron recombination at **11** molecules. Near-white emission was achieved in Device 8b, with maximum quantum efficiency and CIE coordinates being 15.5% and (0.43, 0.43), respectively.

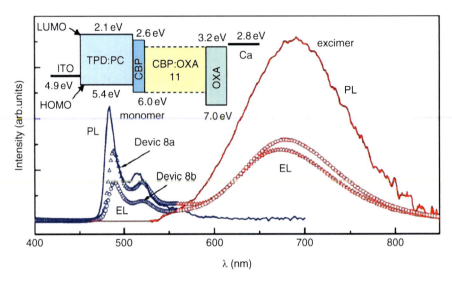

Fig. 7 PL spectra of monomer and excimer emission for **11** and EL spectra of Devices 8a–8b. Reproduced with permission from [50]. Copyright 2007, American Institute of Physics

As the EL spectra of Devices 8a and 8b contain excessive distinctiveness of individual emissive components and did not have green-emitting component, their corresponding CRI values cannot exceed 75. This issue has been addressed by introducing a third emitting component: m-MTDATA for the exciplex emission of **11** in the device. Based on Devices 8a and 8b, Kalinowski et al. fabricated Device 9 with the structure of ITO/TPD:PC (70 nm)/CBP (10 nm)/m-MTDATA:**11** (1:1, 70 nm)/TAZ (70 nm)/Ca. The proposed white-light-generation mechanisms are illustrated in Fig. 8. This device revealed three different emissions (a) molecular excitons of **11** producing phosphorescence, (b) excited homo-molecular dimer of **11** producing excimer emission, and (c) excited hetero-molecular dimer between **11** and m-MTDATA. Device 9 showed η_{ext}, η_p, CIE, and CRI of 6.5%, 9.5 lm W^{-1} (at 500 cd m^{-2}), (0.46, 0.45), and 90, respectively [51]. The EL spectra for Device 9 are shown in Fig. 9.

In 2008, Jabbour et al. reported that **13** shows similar absorption and emission properties to [(F$_2$C^N)Pt(acac)] complexes [52]. The low-energy absorption band at 300–400 nm is assigned to MLCT transition, and the emission at ~480 nm is originated from mixed ^{3}LC-MLCT excited state. The emission quantum efficiency of complex **13** ($\Phi = 0.46$) is much higher than that of [(F$_2$C^N)Pt(acac)] complexes ($\Phi < 0.02$). Blue color OLED had been fabricated using the device structure of ITO/PEDOT:PSS/PVK/26mCPy:1,3-bis[4-*tert*-butylphenyl-1,3,4-oxadiazolyl] phenylene (OXD-7, 49%):**13** (2%)/BCP/CsF/Al with η_{ext}, η_L, and η_p and CIE of 12.6%, 20.6 cd A^{-1}, 6.4 lm W^{-1}, and (0.15, 0.26), respectively (Device 10). Excimer emission of **13** appears with emission maximum at ~650 nm upon increasing the doping concentration of **13**. WOLEDs were obtained in 8% doping concentration with η_{ext}, η_L, and η_p and CIE of 9.3%, 13 cd A^{-1}, 7.3 lm W^{-1} and (0.33, 0.36),

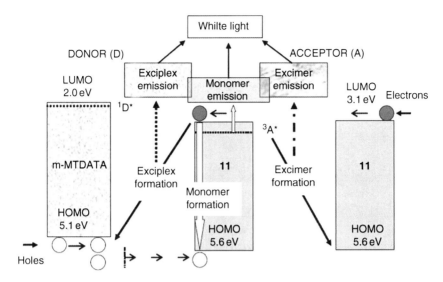

Fig. 8 Proposed generation mechanisms of WOLED based on m-MTDATA acting as an electron-donor (D) to an electron-acceptor (A) molecule of **11** mixed in an emissive layer, D:A (1:1). The monomer **11** triplets (^{3}A*), their combination with ground state **11** molecules [triplet excimers, ^{3}(AA)*], and excited hetero-dimer [^{3}(DA)*] are generated throughout all the emissive layers. Reproduced with permission from [51]. Copyright 2007, WILEY-VCH Verlag GmbH & Co. KGaA, Weinheim

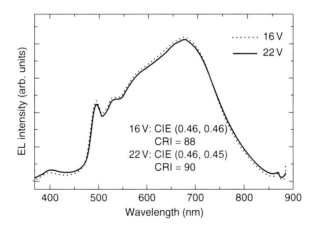

Fig. 9 EL spectra of Device 9 at different operating voltages showing good color stability. Reproduced with permission from [51]. Copyright 2007, WILEY-VCH Verlag GmbH & Co. KGaA, Weinheim

respectively. The EL spectrum of Device 10 is shown in Fig. 10. It is found to be insensitive to the change of the driving voltages, whereas the device brightness increases from 100 to 1,000 cd m^{-2} with changing the voltage from 5.6 to 7.8 V.

Fig. 10 EL spectra of Device 10 at different operating voltages. Reproduced with permission from [54]. Copyright 2008, WILEY-VCH Verlag GmbH & Co. KGaA, Weinheim

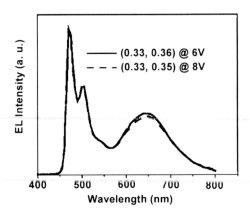

3.2.2 Platinum(II) 3-(6'-(2''-naphthyl)-2'-pyridyl)isoquinolinyl [(RC^N^N)PtCl] Complexes

In 2007, Che et al. reported a series of platinum(II) 3-(6'-(2''-naphthyl)-2'-pyridyl) isoquinolinyl [(RC^N^N)PtCl] complexes (**14–19** – chemical structures are shown in Scheme 7) [33]. These complexes are strongly emissive with emission quantum efficiency of up to 0.68 in dichloromethane solution and have high thermal stability of T_d up to 549 °C. All these complexes show low-energy absorption at 400–470 nm (extinctions between 6,800 and 10,000 dm^3 mol^{-1} cm^{-1}), which is attributed to a ^{1}MLCT: [(5d)Pt → π*(L)] transition, and an absorption tail at around 500 nm, which is tentatively assigned to ^{3}MLCT transition. Strong emissions with λ_{max} at ~530 nm in dichloromethane solution are assigned to have mixed ^{3}IL and ^{3}MLCT parentage.

High-performance orange-emitting OLED has been fabricated using **17** as dopant, and the device characteristics showed η_{ext}, η_L, and η_p of 16.1%, 51.8 cd A^{-1}, and 23.2 lm W^{-1}, respectively [55]. Complexes [(RC^N^N)PtCl] demonstrated the potential in the development of WOLED applications; high-efficiency WOLEDs with dual emissive layers were achieved using a blue-emitting

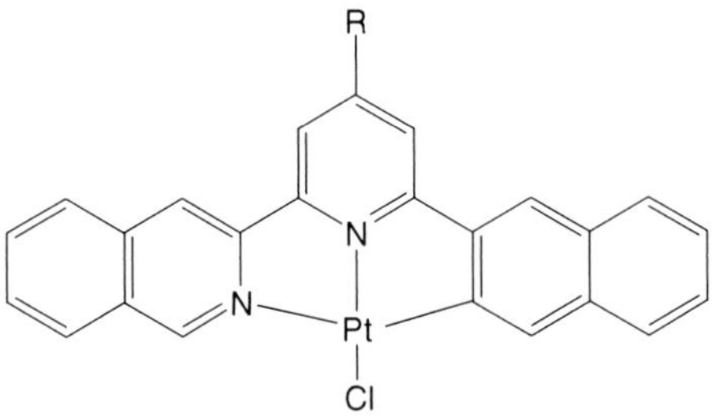

Scheme 7 Chemical structures of complexes **14–19**

14: R = H   **17**: R = 3, 5-^{t}Bu$_2$C$_6$H$_3$
15: R = ^{t}Bu   **18**: R = 3, 5-F$_2$C$_6$H$_3$
16: R = Ph   **19**: R = 3, 5-(CF$_3$)$_2$C$_6$H$_3$

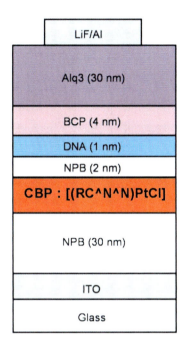

Fig. 11 The device structure of Devices 11 and 12. Reproduced with permission from [54]. Copyright 2007, WILEY-VCH Verlag GmbH & Co. KGaA, Weinheim

fluorescent material (DNA) and phosphorescent [(RC^N^N)PtCl] complexes as an orange or green-yellow guest, doped into CBP host [56]. Figure 11 depicts the device structure of Devices 11 and 12 using dopants **14** and **19**, respectively.

As a 2-nm NPB layer is inserted between CBP:**14/19** and DNA layers, electrons could not approach the CBP:**14/19** layer, and hence the recombination zones for Devices 11 and 12 are confined to the DNA layer. The singlet-state excitons from DNA are responsible for the blue-emitting component, which are transferred to **14/19** through Förster transfer; the triplet excitons are transferred to CBP:**14/19** layer through diffusive transfer which results in yellow-emitting component. The EL spectra of Devices 11 and 12 showed both emissions from DNA and [(RC^N^N)PtCl] with CIE of (0.32, 0.37) and (0.33, 0.34), respectively. Device 11 shows η_{ext}, η_p, and CRI of 11%, 12.6 lm W^{-1}, and 88, respectively, whereas Device 12 demonstrates η_{ext}, η_p, and CRI of 11.8%, 18.4 lm W^{-1}, and 73, respectively (Figs. 12 and 13).

3.3 Tetradentate Platinum(II) Complexes

3.3.1 Platinum(II) Bis(phenoxy)diimine and Schiff Base Complexes [Pt(N$_2$O$_2$)]

In 2003, Che et al. prepared platinum(II) bis(phenoxy)diimiine complexes which show high thermal stability ($T_d > 400\ °C$) and emission quantum efficiency ($\Phi = 0.6$), in which the chemical structures are shown in Scheme 8 [55]. Solution

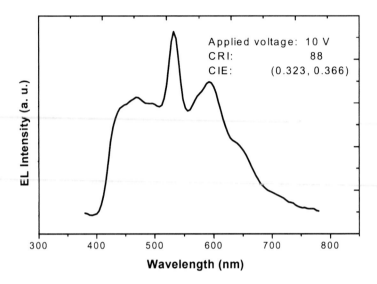

Fig. 12 EL spectrum of Device 11. Reproduced with permission from [54]. Copyright 2007, WILEY-VCH Verlag GmbH & Co. KGaA, Weinheim

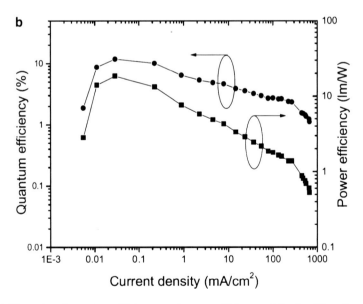

Fig. 13 The external quantum efficiency and power efficiency of Device 12 as functions of current density. Reproduced with permission from [56]. Copyright 2007, WILEY-VCH Verlag GmbH & Co. KGaA, Weinheim

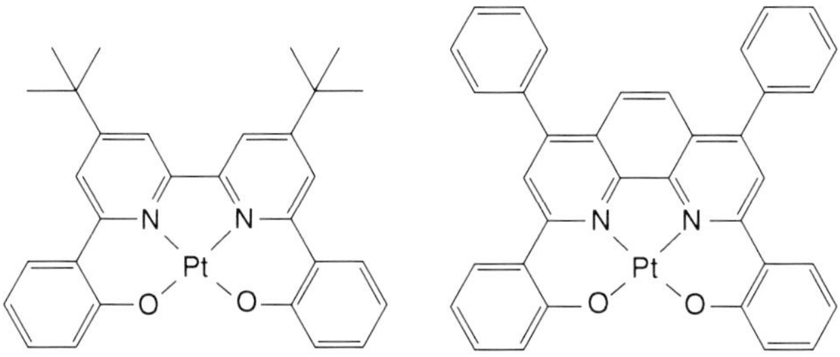

Scheme 8 Chemical structures of complexes **20–21**

emission λ_{max} of **20** and **21** are at 586 and 595 nm which are assigned to mixed excited state with $^3[l - \pi^*(\text{diimine})]$ (l = lone pair/phenoxide) and $^3[\text{Pt}(5d) - \pi^*(\text{diimine})]$ character; the emission quantum yields are up to 0.6 in dichloromethane solution.

High-brightness yellow-emitting OLEDs have been fabricated by using **20** and **21** as emitting materials with the device structure of ITO/NPB/Bepp$_2$:**20** or **21**/LiF/Al (Devices 13a and 14 for **20** and **21**, respectively). In order to fabricate WOLEDs, the doping concentration of **20** had been decreased to give an EL spectrum with two emission peaks at λ_{max} = 453 and 540 nm (Fig. 14). Nevertheless, the CIE coordinates could not fall in the white region, and OLED with greenish yellow (CIE: 0.33, 0.47) electroluminescence was obtained in Device 13c having 0.3% of **20**.

In 2004, another class of emissive [Pt(N$_2$O$_2$)] complexes, namely platinum(II) Schiff base complexes, had been reported, and chemical structures are shown in Scheme 9 [56]. Like platinum(II) bis(phenoxy)diimiine complexes, platinum(II) Schiff base complexes show good thermal stability (T_d up to 406°C) and high emission quantum efficiency (Φ up to 0.19 in acetonitrile solution).

High-performance yellow OLEDs had been fabricated by using **23** as electrophosphorescent dopant, and the maximum device efficiency of 30 cd A^{-1} and 13 lm W^{-1} had been achieved by the configuration of ITO/NPB/CBP:**23**/BCP/Alq$_3$/LiF/Al (Device 15). As the emission bands of **22** and **23** are vibronic structured, the FWHM for EL spectrum of **22** is 30 nm larger than **20** (70 nm for Device 13a, 100 nm for Device 15). Although the emissions, λ_{max}s, of the EL for Devices 13a and 15 are nearly the same (Fig. 15), WOLED with CIE coordinates of (0.33, 0.35) had been obtained by using **22** as yellow-emitting material with the device structure of ITO/NPB (50 nm)/Bepp$_2$:**22** (3.2%, 50 nm)/LiF (1.5 nm)/Al (150 nm) (Device 16).

3.4 Binuclear Platinum(II) Complexes

As platinum(II) complexes adopt square planar geometry and could generate excimer emission upon aggregation, some of the WOLEDs reported in the

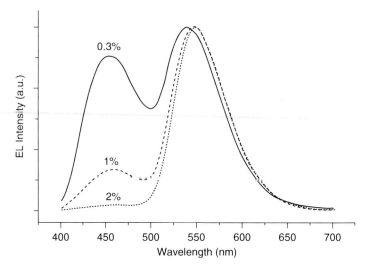

Fig. 14 EL spectra for OLED devices with different concentrations of **20**. (Device 13a: 2%; Device 13b: 1%; and Device 13c: 0.3%) Reproduced with permission from [56]. Copyright 2003, WILEY-VCH Verlag GmbH & Co. KGaA, Weinheim

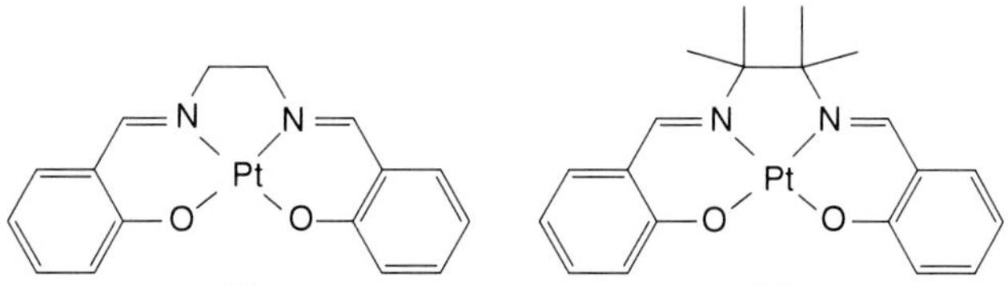

Scheme 9 Chemical structures of complexes **22–23**

literature, making use of the orange to red excimer emission of [(F$_2$C^N)Pt(O^O)] and [Pt(N^C^N)X], are at high dopant concentration. On the other hand, excimer emission can be obtained from dilute solution by putting two platinum(II) complexes in close proximity. We previously demonstrated red emission with λ_{max} at ~650 nm from dilute (5 × 10^{-5} mol dm^{-3}) acetonitrile solution of [Pt$_2$(L)$_2$(μ-dppm)](ClO$_4$)$_2$ (L = 4-(aryl)-6-phenyl-2,2′-bipyridine) complexes. These red emissions show no vibronic structure with their emission; lifetimes and quantum yields are highly sensitive to solvent polarity, but are insensitive to the complex concentration and have been assigned to come from ^{3}[dσ*, π*] (MMLCT) excited state [57].

Thompson et al. prepared a series of binuclear platinum(II) complexes (**24–26** – chemical and crystal structures are shown in Scheme 10) which emit blue, green, and red light depending on the intra- and intermolecular metal–metal distances [58]. The platinum–platinum separation of **24–26** decreases from 3.376 to 2.834 Å

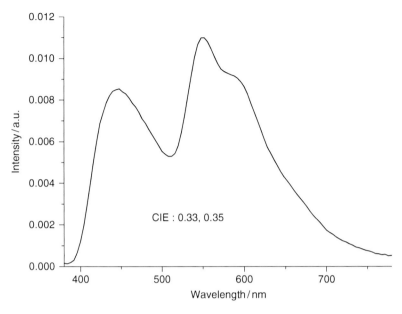

Fig. 15 EL spectrum of Device 16. Reproduced with permission from [56]. Copyright 2004, Royal Society of Chemistry

Scheme 10 Chemical structures and X-ray structures of complexes **24–26**. Reproduced with permission from [58]. Copyright 2006, WILEY-VCH Verlag GmbH & Co. KGaA, Weinheim

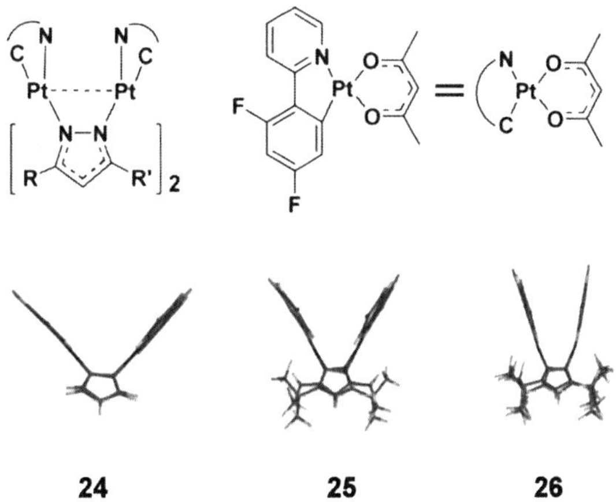

upon increase in the bulkiness of the R and R′ groups. The emission λ_{max} from thin film samples of **24–26** red shifts from 466 to 630 nm. Green (CIE: 0.31, 0.63) and red (0.59, 0.46) OLEDs had been fabricated with **25** and **26**, respectively.

By making use of **24** and **26**, dual-emitting layers (Devices 17–18) and single-emitting layer (Device 19) WOLEDs had been fabricated with the structures shown in Fig. 16. In dual-emitting layer devices, blue light was produced by a layer of mCP doped with 8 wt% **24**, and red light was produced by either a layer of mCP

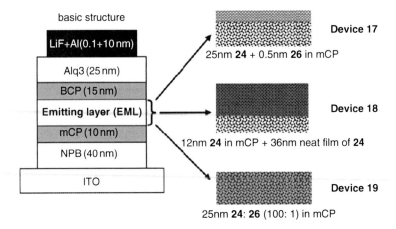

Fig. 16 Device structure of Devices 17–19. Reproduced with permission from [58]. Copyright 2006, WILEY-VCH Verlag GmbH & Co. KGaA, Weinheim

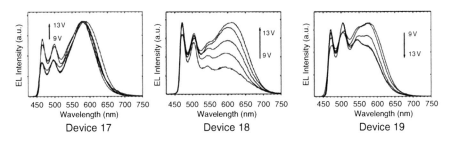

Fig. 17 EL spectra of Devices 17–19 at different operating voltages. Reproduced with permission from [60]. Copyright 2006, WILEY-VCH Verlag GmbH & Co. KGaA, Weinheim

doped with 8 wt% of **26** or a layer of a neat complex **24**. In Device 19, both blue and red lights were produced by codoping complexes **24** and **26** in a mCP layer with 8 wt% of **24**(100):**26**(1) mixture.

Near-white emission had been obtained in Devices 17–19 with CIE of (0.39, 0.45), (0.33, 0.42), and (0.32, 0.48), respectively. However, as the relative emission intensity for the blue- and red-emitting components change with operation voltage in Devices 17–19, these devices did not show good color stability (Fig. 17). The maximum external quantum efficiency of 7.7% has been obtained with Device 17 when operated at 8.5 V.

3.5 Polymeric Materials

Polymeric materials can be solution-processed, and therefore display panels based on light emitting polymeric materials can be fabricated by printing technology.

When compared with devices fabricated by vacuum deposition, the production cost is reduced, the size of the panels can be enlarged, and flexible display can be fabricated by using organic polymeric light-emitting diodes. An important milestone for polymer-based OLED is the announcement by Seiko Epson on a 40-inch OLED display panel which was fabricated by inkjet printer in 2005 SID. Although the development for light-emitting polymer is mainly focused on nonmetal-containing materials such as poly(p-phenylenevinylene) (PPV) and polyfluorene (PFO), the statistical 1:3 singlet to triplet exciton ratio is still a motivation for the development of metal-containing polymers. Among the metal-containing polymers, platinum(II)-containing polymers play an important role in white-emitting polymer OLED.

In 2004, Thompson et al. reported that multifunctional polymers (**27–30**, chemical structures are shown in Scheme 11) containing [(F$_2$C^N)Pt(O^O)] as emitting units show white EL emission [59]. These polymers contain different ratio of triphenylamine as hole-transporting units, oxadiazole as electron-transporting units, and [(F$_2$C^N)Pt(O^O)] as emitting units. By changing the ratio, the EL spectrum changes; the monomer emission of [(F$_2$C^N)Pt(O^O)] unit is dominant at low [(F$_2$C^N)Pt(O^O)] feed ratio, while the excimer emission becomes more noticeable upon increase in the [(F$_2$C^N)Pt(O^O)] feed ratio. The spin-coated device of **27** exhibited near-white emission with CIE of (0.33, 0.50), and the turn on voltage and η_{ext} of the device were 7.8 V and 4.6%, respectively (Device 20).

As discussed in the previous section, singlet host, in combination with triplet-doping materials, becomes the major trend for WOLED fabrication. In 2007, Marder et al. reported platinum(II) polymers containing fluorescent-emitting units [60]. Copolymers with 2,7-di(carbazol-9-yl)fluorene moiety as host material for [(C^N)Pt(O^O)] complexes could be synthesized by using Grubbs' second- or third-generation ruthenium catalysts. When compared with the homo-polymers containing only 2,7-di(carbazol-9-yl)fluorene repeating units, [(C^N)Pt(O^O)]

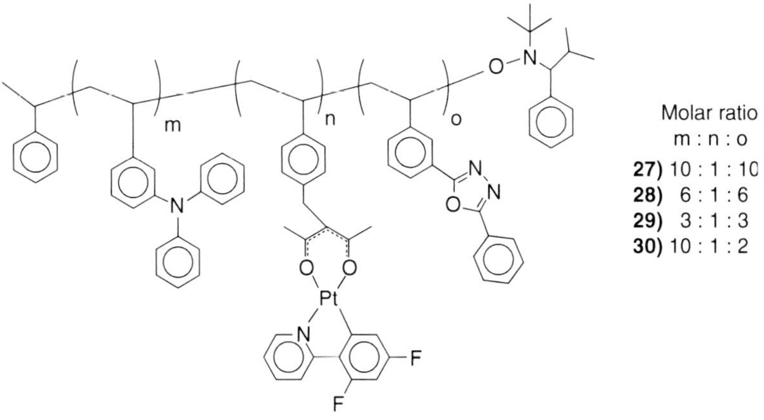

Scheme 11 Chemical structures of polymers **27–30**

moiety-containing polymers have lower molecular weight and broader molecular weight distribution [homo-polymer: M_w: 67,000, PDI: 2.27; copolymers: M_w: ~45,000, PDI: ~3.5 (determine by GPC)]. This has been attributed to the partial decomposition of the ruthenium catalyst. Figure 18 depicts thin-film photoluminescence spectrum recorded with a sample of polymer **32** (chemical structures of **31** and **32** are shown in Scheme 12). The spectrum reveals a near-white broad emission attributed to a combination of higher-energy structured emission from isolated molecules **31**, and a lower-energy unstructured emission from excimer of **31**. Using **32** as the emitting layer, device with structure of ITO/hole transporting polymer/**32**/BCP/Alq$_3$/LiF/Al (Device 21) that exhibited white emission has been fabricated.

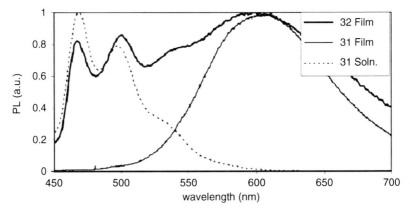

Fig. 18 Photoluminescence spectra for **31** and **32**. Reproduced with permission from [60]. Copyright 2007, American Chemical Society

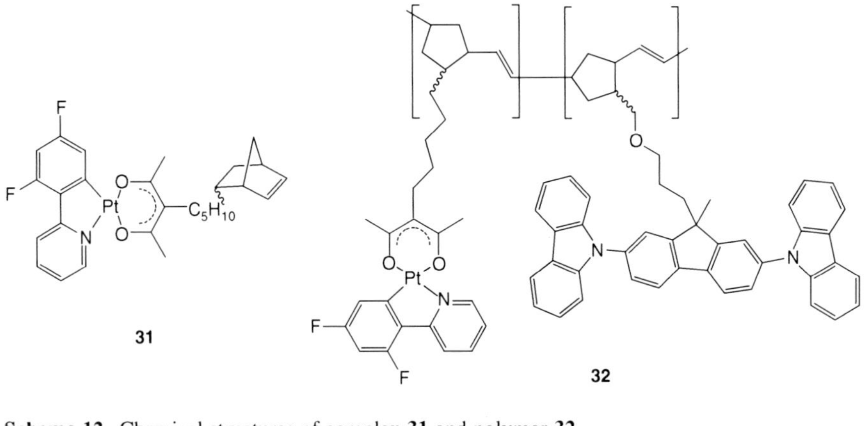

Scheme 12 Chemical structures of complex **31** and polymer **32**

4 Summary

Since WOLEDs exhibit device efficiency that surpasses those in fluorescent tubes, it could become the next generation in the use of lighting appliances. Emitting material is one of the important issues in the development of WOLEDs. Platinum (II)-based materials have shown prominent practical application in OLEDs because they emit from triplet excited states, show high-emission quantum efficiency and possess good thermal stability. Numerous high-efficiency single-color OLEDs had been fabricated by using the above-mentioned [(C^N)Pt(O^O)], [Pt(N^C^N)X], [(RC^N^N)PtCl], and [Pt(N$_2$O$_2$)] complexes. The emissions of these devices cover from blue to near-IR spectral region, which can be widely used for WOLED applications. By making use of complementary colors, high-performance WOLEDs had been fabricated by mixing different platinum(II) complexes with other blue-emitting materials. WOLED fabricated from mononuclear [(F$_2$C^N)Pt(O^O)] was reported to show nearly 100% internal quantum efficiency. WOLEDs with single-emitting layer or single-emitting material were fabricated by using dinuclear platinum(II) complexes, whereas polymeric platinum(II)-based materials offer solution processability for WOLED fabrication.

References

1. Tang CW, Van Slyke SA (1987) Appl Phys Lett 51:913–915
2. Tsujimura T, Kobayashi Y, Murayama K et al (2003) Society for information display. Int Symp Digest 4(1):6–9
3. Rajeswaran G, Itoh M, Boroson M et al (2000) Society for information display. Int Symp Digest 40(1):974–977
4. http://www.oled-display.net/oled-television
5. http://www.lge.co.kr/cokr/product/catalog/FrontProductDetailCmd.laf
6. Kido J, Hongawa K, Okuyama K et al (1994) Appl Phys Lett 64:815–817
7. Kido J, Kimura M, Nagai K (1995) Science 267:1332–1334
8. Park YS, Kang JW, Kang DM et al (2008) Adv Mater 20:1957–1961
9. Cheng G, Zhang Y, Zhao Y et al (2006) Appl Phys Lett 89:043504
10. Kang GW, Ahn YJ, Lim JT et al (2003) Synth Met 137:1029–1030
11. Kim CH, Shinar J (2002) Appl Phys Lett 80:2201–2203
12. Sun Y, Forrest SR (2007) Appl Phys Lett 91:263503
13. Burrows PE, Forrest SR, Sibley SP et al (1996) Appl Phys Lett 69:2959–2961
14. Tokito S, Iijima T, Tsuzuki T et al (2003) Appl Phys Lett 83:2459–2461
15. Lei G, Wang L, Qiu Y (2006) Appl Phys Lett 88:103508
16. Lin MF, Wang L, Wong WK et al (2007) Appl Phys Lett 91:073517
17. Ho CL, Wong WY, Wang Q et al (2008) Adv Funct Mater 18:928–937
18. Su SJ, Gonmori E, Sasabe H et al (2008) Adv Mater 20:4189–4194
19. Kim MS, Jeong CH, Lim JT et al (2008) Thin Solid Films 516:3590–3594
20. Jiang XY, Zhang ZL, Zhang BX et al (2002) Synth Met 129:9–13
21. Chuen CH, Tao YT (2002) Appl Phys Lett 81:4499–4501
22. Zhu FJ, Hua YL, Yin SG et al (2007) J Lumin 122–123:717–719
23. Tao S, Peng Z, Zhang X et al (2006) J Lumin 121:568–572

24. Zheng XY, Zhu WQ, Wu YZ et al (2003) Displays 24:121–124
25. Webb DL, Rossiello LA (1971) Inorg Chem 10:2213–2218
26. Geoffroy GL, Wrighton MS, Hammond GS et al (1974) J Am Chem Soc 96:3105–3108
27. Johnson CE, Eisenberg R, Evans TR et al (1983) J Am Chem Soc 105:1795–1802
28. Ballardini R, Varani G, Indelli MT et al (1986) Inorg Chem 25:3858–3865
29. Barigelletti F, Sandrini D, Maestri M et al (1988) Inorg Chem 27:3644–3647
30. Sandrini D, Maestri M, Balzani V et al (1987) J Am Chem Soc 109:7720–7724
31. Hissler M, Connick WB, Geiger DK et al (2000) Inorg Chem 39:447–457
32. Chan SC, Chan MCW, Wang Y et al (2001) Chem Eur J 7:4180–4190
33. Kui SCF, Sham IHT, Cheung CCC et al (2007) Chem Eur J 13:417–435
34. Donges D, Nagle JK, Yersin H (1997) Inorg Chem 36:3040–3048
35. Donges D, Nagle JK, Yersin H (1997) J Lumin 72–74:658–659
36. Brooks J, Babayan Y, Lamansky S et al (2002) Inorg Chem 41:3055–3066
37. D'Andrade BW, Brooks J, Adamovich V et al (2002) Adv Mater 14:1032–1036
38. Adamovich V, Brooks J, Tamayo A et al (2002) New J Chem 26:1171–1178
39. Williams EL, Haavisto K, Li J et al (2007) Adv Mater 19:197–202
40. He Z, Wong WY, Yu X et al (2006) Inorg Chem 45:10922–10937
41. Bhansali US, Jia H, Lopez MAQ et al (2009) Appl Phys Lett 94:203501
42. Li M, Chen WH, Lin MT et al (2009) Org Electron 10:863–870
43. Ballardini R, Indelli MT, Varani G et al (1978) Inorg Chim Acta 31:L423–L424
44. Bartocci C, Sostero S, Traverso O et al (1980) J Chem Soc Faraday Trans 1 76:797–803
45. Ballardini R, Varani G, Indelli MT et al (1986) Inorg Chem 25:3858–3865
46. Xiang HF, Xu ZX, Roy VAL et al (2008) Appl Phys Lett 92:163305
47. Williams JAG, Beeby A, Davies ES et al (2003) Inorg Chem 42:8609–8611
48. Sotoyama W, Satoh T, Sawatari N et al (2005) Appl Phys Lett 86:153505
49. Cocchi M, Kalinowski J, Virgili D et al (2008) Appl Phys Lett 92:113302
50. Cocchi M, Kalinowski J, Virgili D et al (2007) Appl Phys Lett 90:163508
51. Kalinowski J, Cocchi M, Virgili D et al (2007) Adv Mater 19:4000–4005
52. Yang X, Wang Z, Madakuni S et al (2008) Adv Mater 20:2405–2409
53. Yan BP, Cheung CCC, Kui SCF et al (2007) Appl Phys Lett 91:063508
54. Yan BP, Cheung CCC, Kui SCF et al (2007) Adv Mater 19:3599–3603
55. Lin YY, Chan SC, Chan MCW et al (2003) Chem Eur J 9:1264–1272
56. Che CM, Chan SC, Xiang HF et al (2004) Chem Commun 1484–1485
57. Lai SW, Chan MCW, Cheung TC et al (1999) Inorg Chem 38:4046–4055
58. Ma B, Djurovich PI, Garon S et al (2006) Adv Funct Mater 16:2438–2446
59. Furuta PT, Deng L, Garon S et al (2004) J Am Chem Soc 126:15388–15389
60. Cho JY, Domercq B, Barlow S et al (2007) Organometallics 26:4816–4829

Solid-State Light-Emitting Electrochemical Cells Based on Cationic Transition Metal Complexes for White Light Generation

Hai-Ching Su, Ken-Tsung Wong, and Chung-Chih Wu

Contents

1 Introduction 106
   1.1 Features of Light-Emitting Electrochemical Cells 106
   1.2 Organization of This Chapter 106
2 Review of Light-Emitting Electrochemical Cells Based on Cationic Transition Metal Complexes 107
   2.1 Increasing Device Efficiency 107
   2.2 Color Tuning 118
   2.3 Lengthening Device Lifetime 120
   2.4 Shortening Turn-On Time 121
3 White Light-Emitting Electrochemical Cells Based on Cationic Transition Metal Complexes 127
   3.1 Photophysical and Electrochemical Properties of Novel Blue-Green and Red-Emitting Cationic Iridium Complexes 127
   3.2 Electroluminescent Properties of White Light-Emitting Electrochemical Cells Based on Host–Guest Cationic Iridium Complexes 131
4 Outlook 133
References 134

Abstract Solid-state light-emitting electrochemical cells (LECs) based on cationic transition metal complexes (CTMCs) exhibit several advantages over conventional light-emitting diodes such as simple fabrication processes, low-voltage operation,

---

H.-C. Su (✉)
Institute of Lighting and Energy Photonics, National Chiao Tung University, No. 301, Gaofa 3rd Road, Guiren Township, Tainan County 711, Taiwan, ROC
e-mail: haichingsu@mail.nctu.edu.tw

K.-T. Wong
Department of Chemistry, National Taiwan University, No. 1, Sec. 4, Roosevelt Road, Taipei 106, Taiwan, ROC

C.-C. Wu (✉)
Department of Electrical Engineering, National Taiwan University, No. 1, Sec. 4, Roosevelt Road, Taipei 106, Taiwan, ROC
e-mail: chungwu@cc.ee.ntu.edu.tw

and high power efficiency. Hence, white CTMC-based LECs may be competitive for lighting applications. In this chapter, we review previous important works on CTMC-based LECs, such as increasing device efficiency, color tuning, lengthening device lifetime, and shortening turn-on time. Our demonstration of white CTMC-based LECs by using the host–guest strategy is then described.

1 Introduction

1.1 Features of Light-Emitting Electrochemical Cells

White organic light-emitting diodes (OLEDs) based on polymers and small-molecule materials have attracted much attention because of their potential applications in flat-panel displays and solid-state lighting [1–5]. Compared with conventional white OLEDs [6, 7], solid-state white light-emitting electrochemical cells (LECs) [8–10] possess several promising advantages. LECs generally require only a single emissive layer, which can be easily processed from solutions, and can conveniently use air-stable electrodes. The emissive layer of LECs contains mobile ions, which can drift toward electrodes under an applied bias. The spatially separated ions induce doping (oxidation and reduction) of the emissive materials near the electrodes, that is, p-type doping near the anode and n-type doping near the cathode [8–10]. The doped regions induce ohmic contacts with the electrodes and consequently facilitate the injection of both holes and electrons, which recombines at the junction between p- and n-type regions. As a result, a single-layered LEC device can be operated at very low voltages (close to E_g/e, where E_g is the energy gap of the emissive material and e is elementary charge) with balanced carrier injection, giving high power efficiencies. Furthermore, air-stable metals, for example, gold and silver, can be used since carrier injection in LECs is relatively insensitive to work functions of electrodes. Therefore, the power-efficient properties and easy fabrication processes make LECs competitive in solid-state lighting technologies.

In the past few years, cationic transition metal complexes (CTMCs) have also been increasingly studied for solid-state LECs because of the following several advantages over conventional LECs or polymer LECs (a) ion-conducting material is not needed since these metal complexes are intrinsically ionic and (b) higher electroluminescent (EL) efficiencies could be achieved due to the phosphorescent nature of the transition metal complexes. The development of solid-state LECs based on CTMCs thus will be described and discussed in this chapter.

1.2 Organization of This Chapter

This chapter is organized as follows. Section 2 reviews advances of LECs based on CTMCs, including topics of device efficiency (Sect. 2.1), colors (Sect. 2.2), device lifetimes (Sect. 2.3), and turn-on times (Sect. 2.4). Section 3 is devoted to white LECs based on CTMCs. In Sect. 3.1, photophysical and electrochemical properties of newly developed blue-green and red emitting cationic iridium complexes are discussed. EL properties of white LECs based on host–guest cationic iridium complexes are then discussed in Sect. 3.2. Finally, we give an outlook about white LECs based on CTMCs.

2 Review of Light-Emitting Electrochemical Cells Based on Cationic Transition Metal Complexes

The first solid-state LEC was demonstrated with a polymer blend sandwiched between two electrodes [8]. The polymer blend was composed of an emissive conjugated polymer, a lithium salt (lithium trifluoromethanesulfonate), and an ion-conducting polymer (polyethylene oxide). The salt provides mobile ions and the ion-conducting polymer prevents this blend film from phase separation that may be induced by polarity discrepancy between the conjugated polymer and the salt. More recently, CTMCs have also been used in LECs, which show several advantages over conventional polymer LECs. In such devices, no ion-conducting material is needed since these metal complexes are intrinsically ionic. They generally show good thermal stability and charge-transport properties. Furthermore, high EL efficiencies are expected due to the phosphorescent nature of these metal complexes.

2.1 Increasing Device Efficiency

The first solid-state LEC based on transition metal complexes was reported in 1996 [11], where a ruthenium poly-pyridyl complex was utilized as the emissive material. Since then, many efforts have been made to improve performances of the LECs. In 1999, LECs based on low-molecular-weight ruthenium complexes were reported to have external quantum efficiency (EQE), which is defined as photons/electrons, of 1% [12]. In 2000, LECs using $[Ru(bpy)_3](ClO_4)_2$ (where bpy is 2,2′-bipyridine) with the gallium–indium eutectic electrode showed an EQE up to 1.8% [13]. Later, $[Ru(bpy)_3](PF_6)_2$ blended with poly(methyl methacrylate) (PMMA) was reported to improve the film quality and increase the EQE to 3% [14]. In 2002, a single-crystal LEC was made by repeatedly filling nearly saturated solution of $[Ru(bpy)_3](ClO_4)_2$ between two ITO (indium tin oxide) slides, followed by evaporating the solvent [15]. Such devices possessed low turn-on voltages and exhibited an EQE of 3.4%. Further improvement of performances was achieved by reducing self-quenching of excited states in $[Ru(bpy)_3]^{2+}$ with adding alkyl substituents on the bpy ligands [16], raising the EQE to 4.8% under the DC bias and to 5.5% under the pulsed driving. Yet a further improvement of the $[Ru(bpy)_3](ClO_4)_2$ device was achieved by forming a heterostructure device, thus moving the emission zone away from the electrode and giving an efficiency of 6.4% [17].

More efficient cationic iridium complexes have also been used in LECs. In 2004, LECs based on the yellow-emitting (560 nm) cationic iridium complex [Ir(ppy)$_2$(dtb-bpy)]PF$_6$, where ppy is phenylpyridine and dtb-bpy is 4,4'-di-*tert*-butyl-2,2'-bipyridine, were reported, exhibiting efficiencies of 5% and 10 lm W^{-1} [18, 19]. On the one hand, replacing the ppy ligands by F-mppy ones, where F-mppy is 2-(4'-fluorophenyl)-5-methylpyridine, led to green emission (531 nm) and an EL efficiency of 1.8% [19, 20]. On the other hand, employing the dF(CF$_3$)ppy ligands, where dF(CF$_3$)ppy is 2-(2,4-difluorophenyl)-5-trifluoromethylpyridine, increased the energy gap of the cationic iridium complexes and shifted the EL emission to blue-green (500 nm, with an EQE of 0.75%) [21]. Another series of cationic phenylpyrazole-based iridium complexes were also recently reported to give blue (492 nm), green (542 nm), and red (635 nm) emission [22]. Blue, green, and red devices made of these complexes on poly(3,4-ethylenedioxythiophene):poly(styrene sulfonate) (PEDOT:PSS) coated ITO showed EQEs of 4.7%, 6.9%, and 7.4%, respectively.

Most LEC devices described above were made in a single-layered neat-film structure. In a neat film of an emissive material, interactions between closely packed molecules usually lead to quenching of excited states, detrimental to EQEs of devices. Thus, in addition to tuning of emission colors, the capability of the ligands to provide steric hindrance for suppressing self-quenching must also be carefully considered in designing ligands for emissive cationic metal complexes. In our previous studies of oligofluorenes, we had found that the introduction of aryl substitutions onto the tetrahedral C9 of oligofluorenes can provide effective hindrance to suppress interchromophore packing and self-quenching yet without perturbing energy gaps of the molecules [23–27], making the photoluminescent quantum yields (PLQYs) of oligo(9,9-diarylfluorene)s in neat films rather close to those in dilute solutions. Thus, we introduced 4,5-diaza-9,9'-spirobifluorene (SB) [28] as a steric and bulky auxiliary ligand of cationic iridium complexes, [Ir(ppy)$_2$(SB)]PF$_6$ (**1**) and [Ir(dFppy)$_2$(SB)]PF$_6$ (**2**) (Fig. 1), and investigate the influence of the SB ligand on reducing the self-quenching [29].

Spin-coated neat films of **1** and **2** exhibit PL spectra similar to those observed for their acetonitrile (MeCN) solutions (Fig. 2). However, spin-coated neat films of **1** and **2** show higher PLQYs (0.316 for **1**, 0.310 for **2**) and longer excited-state lifetimes (0.60 µs for **1**, 0.59 µs for **2**) than their MeCN solutions (Table 1). Since MeCN is a strongly polar solvent and the emitters are ionic, the photophysical properties of complexes **1** and **2** perhaps are significantly perturbed by strong solute–solvent interaction, rendering the observed PLQYs and lifetimes of **1** and **2** in MeCN less intrinsic. This indeed can be verified by measuring photophysical properties of **1** and **2** in a less polar solvent (yet still with enough solubility for spectroscopic measurements), such as dichloromethane (DCM). In DCM (5×10^{-5} M), both **1** and **2** exhibit longer excited-state lifetimes (0.79 µs for **1**, 0.42 µs for **2**) and higher PLQYs (0.467 for **1**, 0.364 for **2**) than in MeCN (Table 1).

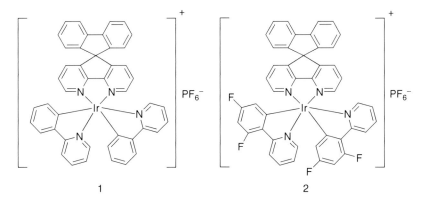

Fig. 1 Molecular structures of complexes **1** and **2**

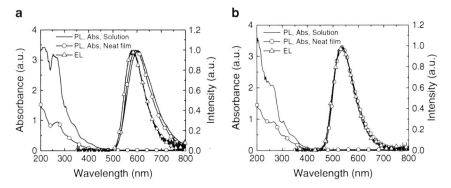

Fig. 2 Absorption and PL spectra in acetonitrile solutions and in neat films of (**a**) complex **1** and (**b**) complex **2**. Also shown in (**a**) and (**b**) are EL spectra of complex **1** and complex **2**, respectively. Reproduced with permission from [29]. Copyright 2007, WILEY-VCH Verlag GmbH & Co. KGaA, Weinheim

Thus, to better characterize the intrinsic photophysical properties of complexes **1** and **2** at the room temperature, complexes **1** and **2** were dispersed (with 1.5 mol%) in a large-gap thin-film host of m-bis(N-carbazolyl)benzene (mCP), which is rather nonpolar and has a large triplet energy. The measured PLQYs and excited-state lifetimes of the dispersed mCP films are (0.667, 0.81 μs) for **1** and (0.421, 0.74 μs) for **2** (Table 1). Thus **1** and **2** dilutely dispersed in mCP exhibit higher PLQYs and longer excited-state lifetimes than those in neat films. Shorter lifetimes in neat films indicate that interaction between closely packed molecules provides additional deactivation pathways, shortening excited-state lifetimes, and lowering the PLQYs. However, PLQYs of complexes **1** and **2** in neat films still retain ~50% and ~75% of those in mCP blend films. These are indeed rather high retaining percentages for PLQYs in neat films when compared with other phosphorescent molecules. For instance, [Ir(ppy)$_3$] [1.5 mol% dispersed in 4,4' bis(carbazol-9-yl)

Table 1 Summary of physical properties of complexes **1** and **2**

Complex	$\lambda_{max, PL}$ (nm)[a]		PLQY, lifetime (μs)[b]		$E^{ox}_{1/2}$ (V)[c]	$E^{red}_{1/2}$ (V)[d]	$\Delta E_{1/2}$ (V)[e]
	Solution[f]	Neat film	Solution	Film			
1	605	593	0.226, 0.33[f] 0.467, 0.79[h] –, 4.31[j] –, 3.27[l]	0.316, 0.60[g] 0.667, 0.81[i] 0.381, 0.72[k]	1.35	–1.33	2.64
2	535	535	0.278, 0.39[f] 0.364, 0.42[h] –, 4.49[j] –, 4.29[l]	0.310, 0.59[g] 0.421, 0.74[i] 0.329, 0.55[k]	1.70	1.26	2.92

[a]PL peak wavelength
[b]Photoluminescence quantum yields and the excited-state lifetimes
[c]Oxidation potential
[d]Reduction potential
[e]The electrochemical gap $\Delta E_{1/2}$ is the difference between $E^{ox}_{1/2}$ and $E^{red}_{1/2}$ corrected by potentials of the ferrocenium/ferrocene redox couple
[f]Measured in acetonitrile (5×10^{-5} M) at room temperature
[g]Neat films
[h]Measured in dichloromethane (5×10^{-5} M) at room temperature
[i]**1** and **2** were dispersed (1.5 mol%) in mCP films
[j]Measured in acetonitrile (5×10^{-5} M) at 77 K
[k]Films with 0.75 mol [BMIM][PF$_6$] per mole of **1** and **2**
[l]Measured in dichloromethane (5×10^{-5} M) at 77 K
Reproduced with permission from [29]. Copyright 2007, WILEY-VCH Verlag GmbH & Co. KGaA, Weinheim

biphenyl (CBP)] and bis[(4,6-difluorophenyl)pyridinato-N,C^2](picolinato)iridium (III) [FIrpic] (1.4 mol% dispersed in mCP) films had been reported to have very high PLQYs of 97 ± 2% and 99 ± 1%, respectively [30], while PLQYs of [Ir(ppy)$_3$)] and [FIrpic] neat films are only ~3% and ~15%, respectively. On the one hand, severe self-quenching in [Ir(ppy)$_3$] and [FIrpic] implies that the ppy, dFppy, and picolinic acid ligands cannot provide enough hindrance against intermolecular interactions in neat films. On the other hand, the high retaining percentages in neat-film PLQYs of complex **1** (with two ppy ligands and one SB ligand) and complex **2** (with two dFppy ligands and one SB ligand) compared with those in dispersed films indicate that SB ligands in these compounds provide effective steric hindrance and greatly reduce self-quenching. It is noted that self-quenching in neat films of [FIrpic] is not as severe as that in neat films of [Ir(ppy)$_3$]. It is likely that the fluoro substituents on the ppy ligands somewhat hinder the intermolecular interactions [30]. A similar effect is also observed here for complex **2**, which exhibits a higher PLQY retaining percentage (74%) than complex **1**.

In the operation of LEC devices, when a constant bias voltage is applied, a delayed EL response that is associated with the time needed for counterions in the LECs to redistribute under a bias is typically observed. For the cases of neat films of complexes **1** and **2**, the redistribution of the anions (PF$_6^-$) leads to the formation of a region of Ir(IV)/Ir(III) complexes (p-type) near the anode and a region of Ir(III)/Ir

(II) complexes (n-type) near the cathode [31]. With the formation of p- and n-regions near the electrodes, carrier injection is enhanced, leading to a gradually increasing device current and emission intensity. Devices based on neat films of complexes **1** and **2** (with the structure of glass substrate/ITO/complex **1** or **2** (100 nm)/ Ag) exhibited very long response times. Very slow device response (e.g., tens of hours to reach the maximum brightness) had also been observed before for other LEC materials (e.g., [Ru(bpy)$_3$](PF$_6$)$_2$ derivatives with esterified ligands) [12]. Plausibly, bulky side groups on molecules impede the migration of ions. Thus to accelerate formation of the p- and n-doped regions in the emissive layer, 0.75 mol ionic liquid 1-butyl-3-methylimidazolium hexafluorophosphate [BMIM][PF$_6$] per mole of complex **1** or **2** was added to provide additional anions (PF$_6^-$) [32].

It had been reported that in polymer LECs, incorporation of polar salts into conjugated polymer films might induce aggregates or phase separation due to discrepancy in polarity [33–35]. Thus, to study the effect of the [BMIM][PF$_6$] addition on thin-film morphologies of complexes **1** and **2**, atomic force microscopy (AFM) of thin films was performed. As shown in Fig. 3, the AFM micrographs for films of complexes **1** and **2** with and without [BMIM][PF$_6$] coated on ITO glass substrates show no significant differences and all give similar root-mean-square (RMS) roughness of ~1 nm. At this concentration of [BMIM][PF$_6$] in complex **1** or **2** (0.75:1, molar ratio), no particular features of aggregation or phase separation were

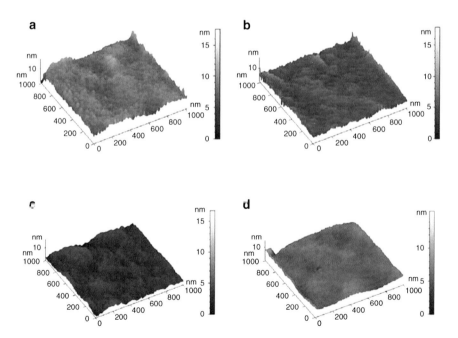

Fig. 3 3D AFM micrographs of (**a**) neat film of complex **1**, (**b**) blend film of [BMIM][PF$_6$] and complex **1** (0.75:1, molar ratio), (**c**) neat film of complex **2**, and (**d**) blend film of [BMIM][PF$_6$] and complex **2** (0.75:1, molar ratio). Reproduced with permission from [29]. Copyright 2007, WILEY-VCH Verlag GmbH & Co. KGaA, Weinheim

observed, and uniform spin-coated thin films could be routinely obtained. Such characteristics may be associated with the ionic nature of complexes **1** and **2**, which may make them more compatible with the added salts. Further, to examine the effects of the [BMIM][PF$_6$] addition on photophysical properties of complexes **1** and **2**, PLQYs and excited-state lifetimes of blend films were also measured and are shown in Table 1. In general, films of both **1** and **2** containing [BMIM][PF$_6$] (1:0.75, molar ratio) exhibit PLQYs and excited-state lifetimes comparable to those of neat films, indicating that the addition of [BMIM][PF$_6$] does not induce particular quenching of emission. Stable operation was also achieved in devices using such a formulation. In the following, device characteristics based on the structure of [glass substrate/ITO/**1**:[BMIM][PF$_6$] or **2**:[BMIM][PF$_6$] (100 nm)/Ag] are discussed and are summarized in Table 2.

A distinct characteristic of LECs is that they can be operated under a bias voltage close to E_g/e. As shown in Table 1, the electrochemical gaps ($\Delta E_{1/2}$), which were derived from the difference between $E_{1/2}^{ox}$ and $E_{1/2}^{red}$ corrected with potentials of the ferrocenium/ferrocene redox couple, for complexes **1** and **2** are 2.64 eV and 2.92 eV, respectively. The devices based on complexes **1** and **2** were thus first tested under the biases of 2.6 V and 2.9 V, respectively, although the energy gaps in films are usually smaller than those in solutions because of the environmental polarization. EL spectra of the devices based on complexes **1** and **2** (added with [BMIM][PF$_6$] are shown in Figs. 2a and 2b, respectively, for comparison with their PL spectra. EL spectra are basically similar to PL spectra, indicating similar emission mechanisms. Commission internationale de l'Eclairage (CIE) coordinates for the EL spectra of complexes **1** and **2** are (0.51, 0.48) and (0.35, 0.57), respectively. Time-dependent brightnesses and current densities of the devices operated under bias conditions described above are shown in Figs. 4a and 4b for complexes **1** and **2**, respectively. Both devices exhibited similar electrical characteristics. On the one hand, the currents of both devices first increased with time after the bias was applied and then stayed at a constant level after 350–400 min. On the other hand,

Table 2 Summary of the LEC device characteristics based on complexes **1** and **2**

Complex	Bias (V)	$\lambda_{max,\ EL}$ (nm)[a]	t_{max} (min)[b]	L_{max} cd m^{-2}	$\eta_{ext,\ max}$, $\eta_{p,\ max}$ (%, lm W^{-1})[d]	Lifetime (h)[e]
1	2.6	580	170	330	6.2, 19.0	26
	2.5		150	100	7.1, 22.6	54
2	2.9	535	85	145	6.6, 23.6	6.7
	2.8		90	52	7.1, 26.2	12

[a]EL peak wavelength
[b]Time required to reach the maximal brightness
[c]Maximal brightness achieved at a constant bias voltage
[d]Maximal external quantum efficiency and maximal power efficiency achieved at a constant bias voltage
[e]The time for the brightness of the device to decay from the maximum to half of the maximum under a constant bias voltage
Reproduced with permission from [29]. Copyright 2007, WILEY-VCH Verlag GmbH & Co. KGaA, Weinheim

Solid-State Light-Emitting Electrochemical Cells   113

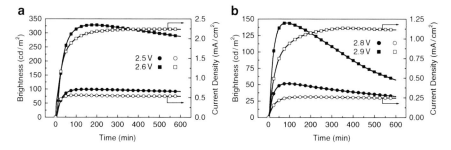

Fig. 4 The time-dependent brightness and current density of the single-layered LEC device for (**a**) complex **1** driven at 2.6 or 2.5 V and (**b**) complex **2** driven at 2.9 or 2.8 V. Reproduced with permission from [29]. Copyright 2007, WILEY-VCH Verlag GmbH & Co. KGaA, Weinheim

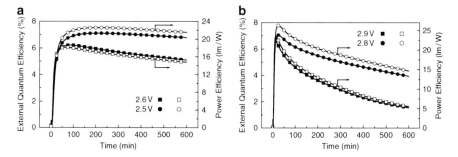

Fig. 5 The time-dependent EQE and the corresponding power efficiency of the single-layered LEC device for (**a**) complex **1** driven at 2.6 or 2.5 V and (**b**) complex **2** driven at 2.9 or 2.8 V. Reproduced with permission from [29]. Copyright 2007, WILEY-VCH Verlag GmbH & Co. KGaA, Weinheim

the brightness first increased with the current and reached the maximum of 330 cd m^{-2} for complex **1** at ~170 min and of 145 cd m^{-2} for complex **2** at ~85 min. The brightness then decreased with time even though the device current stayed rather constant. The decrease in brightness was irreversible, that is, the maximum brightness obtained in the first measurement could not be fully recovered in the followed measurements even under the same driving conditions. It is rationally associated with the degradation of the emissive material during the LEC operation, which was commonly seen in LEC devices [36].

Time-dependent EQEs and corresponding power efficiencies of the complex **1** device under the 2.6-V bias and the complex **2** device under the 2.9-V bias are shown in Figs. 5a and 5b, respectively. Both devices exhibited similar time evolution in EQE. When a forward bias was just applied, the EQE was rather low due to unbalanced carrier injection. During the formation of the p- and n-type regions near electrodes, the balance of the carrier injection was improved and the EQE of the device thus increased rapidly. The peak EQE and the peak power efficiency are

(6.2%, 19.0 lm W^{-1}) for the complex **1** device under the 2.6-V bias and (6.6%, 23.6 lm W^{-1}) for the complex **2** device under the 2.9-V bias. One notices that the peak efficiencies occurred before the currents reached their final maximal values. Such a phenomenon may be associated with two factors. First, although with the formation of the p- and n-type regions near electrodes, both contacts are becoming more ohmic and the carrier injection at both electrodes are becoming more balanced; however, the carrier recombination zone may consequently move toward one of the electrodes because of discrepancy in electron and hole mobilities. The recombination zone moving to the vicinity of an electrode may cause exciton quenching such that the EQE of the device would decrease with time while the current and the brightness are still increasing. Besides, degradation of the emissive material under a high field would also contribute to the decrease in the EQE when the recombination zone is still moving or when the recombination zone is fixed.

Both LEC devices driven under 0.1-V lower biases exhibit higher peak EQEs and lower degradation rates. As shown in Figs. 5a and 5b, the peak EQE and the peak power efficiency are (7.1%, 22.6 lm W^{-1}) for the complex **1** device under the 2.5-V bias and (7.1%, 26.2 lm W^{-1}) for the complex **2** device under the 2.8-V bias. It is interesting to note that the peak EQEs of the single-layered devices (no PEDOT:PSS is used) based on complexes **1** and **2** approximately approach the upper limits that one would expect from the PLQYs of their neat films, when considering an optical out-coupling efficiency of ~20% from a typical layered light-emitting device structure. To our knowledge, these EQEs are among the highest values reported for orange-red (or yellow) and green solid-state LECs based on CTMCs. Such results indicate that CTMCs with superior steric hindrance are essential and useful for achieving highly efficient solid-state LECs.

Since these newly developed cationic complexes (**1** and **2**) are intrinsically efficient, to further reduce self-quenching and increase EL efficiency, one possible approach is to spatially disperse the emitting complex (guest) into a matrix complex (host), as previously reported for conventional OLEDs and solid-state LECs [37–39]. The mixed host–guest films (~100 nm) for PL studies were spin-coated onto quartz substrates using mixed solutions of various ratios. Since in LECs, a salt [BMIM][PF$_6$] of 19 wt% (where BMIM is 1-butyl-3-methylimidazolium) was also added to provide additional mobile ions and to shorten the device response time [29, 32], PL properties of the host–guest–salt three-component system were also characterized.

Figure 6a shows the absorption spectrum of the guest and the PL spectra of the host–guest two-component systems having various guest concentrations. With the increase of the guest concentration, a gradual red shift from the host-like emission to the guest-like emission is observed. As shown in Fig. 6b, the highest PLQY of ~37% (vs. ~31% of neat host and guest films) is obtained at the guest concentration of 25 wt%, at which the emission is nearly completely from the guest. Accompanying this enhanced PLQY is the longer excited-state lifetime (0.69 μs) when compared with those of neat films (0.59 μs for the host and 0.60 μs for the guest), indicating the effectiveness of the dispersion in suppressing quenching mechanisms of guest molecules. Less complete transfer is observed at lower guest concentrations,

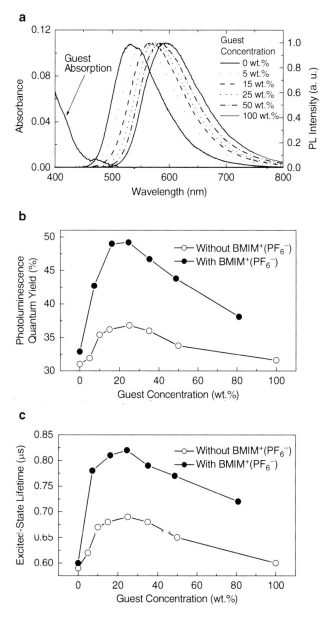

Fig. 6 (a) The absorption spectrum of the neat guest film and PL spectra of host–guest films with various guest concentrations (without [BMIM][PF$_6$]). (b) Photoluminescence quantum yields and (c) excited-state lifetimes as a function of the guest concentration for host–guest films without and with [BMIM][PF$_6$] (19 wt%). Reproduced with permission from [49]. Copyright 2006, American Institute of Physics

which may be associated with the less extensive overlap between the host emission and the weak guest absorption. A calculation of the Förster radius for the host–guest energy transfer gives a small value of <16 Å [40] indicating inefficient energy transfer at lower guest concentrations. With the addition of [BMIM][PF$_6$] (19 wt%), the trend in PL properties [spectra (not shown), PLQYs (Fig. 6b), and lifetimes (Fig. 6c)] as a function of the guest concentration is similar to those of the two-component system. Yet, an even higher PLQY of ~50% (about 1.6× enhancement compared with PLQYs of neat host and guest films) and longer excited-state lifetime of ~0.82 μs are observed around the guest concentration of 25 wt%. It appears that [BMIM][PF$_6$] not only provides additional mobile ions but is also effective in suppressing interchromophore quenching.

Figure 7a shows the time-dependent brightness and current density under constant biases of 2.5–2.7 V for the LEC using the mixture giving the highest PLQY (i.e., with host, guest, and [BMIM][PF$_6$] concentrations of 56 wt%, 25 wt%, and 19 wt%, respectively). After the bias was applied, on the one hand, the current first increased and then stayed rather constant. On the other hand, the brightness first increased with the current and reached the maxima of 10, 30, and 75 cd m^{-2} at <1 h under biases of 2.5 V, 2.6 V, and 2.7 V, respectively. The brightness then dropped with time with a rate significantly depending on the bias voltage (or current). Corresponding time-dependent EQEs and power efficiencies of the same device are shown in Fig. 7b. When a forward bias was just applied, the EQE was rather low due to poor carrier injection. During the formation of the p- and n-type regions near electrodes, the capability of carrier injection was improved and the EQE thus rose rapidly. The peak EQE, current, and peak power efficiencies at 2.5 V, 2.6 V, and 2.7 V are (10.4%, 29.3 cd A^{-1}, 36.8 lm W^{-1}), (9.9%, 27.9 cd A^{-1}, 33.7 lm W^{-1}), and (9.4%, 26.5 cd A^{-1}, 30.8 lm W^{-1}), respectively.

The maximum current density vs. voltage characteristics of LECs with various guest concentrations are shown in Fig. 8a. The bias voltage required for same

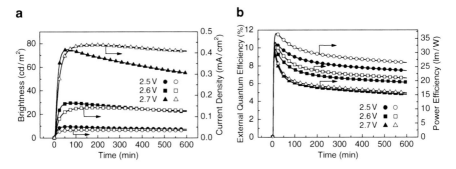

Fig. 7 (a) Brightness (*solid symbols*) and current density (*open symbols*) and (b) external quantum efficiency (*solid symbols*) and power efficiency (*open symbols*) as a function of time under a constant bias voltage of 2.5–2.7 V for the host–guest LEC with host, guest and [BMIM][PF$_6$] concentrations of 56 wt%, 25 wt%, and 19 wt%, respectively. Reproduced with permission from [49]. Copyright 2006, American Institute of Physics

Solid-State Light-Emitting Electrochemical Cells 117

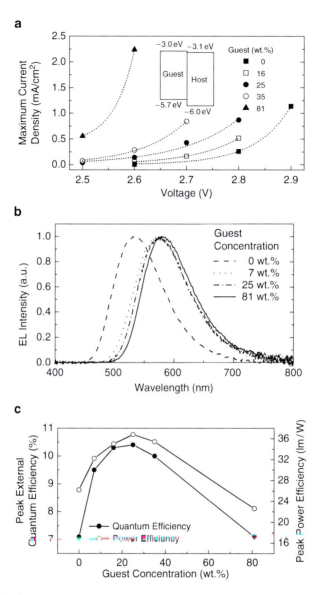

Fig. 8 (**a**) Maximum current density vs. voltage characteristics for LECs with various guest concentrations. (**b**) EL spectra (at 2.8 V) for LECs with various guest concentrations. (**c**) Peak external quantum efficiencies and peak power efficiencies (at current densities <0.1 mA cm^{-2}) of host–guest LECs as a function of the guest concentration. *Inset* of (**a**): the energy level diagram of the host and guest molecules. Reproduced with permission from [49]. Copyright 2006, American Institute of Physics

current drops as the guest concentration is raised. This may be understood by the energy level diagram obtained from cyclic voltammetry [29] (inset of Fig. 8a). Such an energy level alignment favors carrier injection/transport (at least for holes) through the smaller-gap guest and direct carrier recombination/exciton formation on the guest (rather than host–guest energy transfer) if the guest concentration is high enough. EL spectra of host–guest LECs with various guest concentrations (Fig. 8b) indeed support the mechanism of direct exciton formation on guest molecules, which would greatly reduce host emission due to incomplete energy transfer. Compared with PL spectra (e.g., Fig. 6), EL spectra are much less dependent on the guest concentration. Even with a guest concentration as low as 7 wt%, the EL spectrum is almost the same as that of the pure guest device [i.e., the LEC with 81 wt% of guest and 19 wt% of [BMIM][PF$_6$]], indicating predominant guest emission.

Figure 8c shows peak EQEs and peak power efficiencies for host–guest LECs as a function of the guest concentration (at current densities <0.1 mA cm^{-2}). The peak EQEs and power efficiencies roughly follow the trend of the PLQYs. The highest peak EQE and power efficiency of 10.4% and 36.8 lm W^{-1} are achieved at the guest concentration of 25 wt%, coincident with the concentration giving the highest PL efficiency. Such an EQE represents a 1.5× enhancement compared with those of pure host and guest devices. It is also worth noting that such a peak EQE approximately approaches the upper limit that one would expect from the PLQY of the host–guest emissive layer (~50%), when considering an optical out-coupling efficiency of ~20% from a typical layered light-emitting device structure. Such EQEs and power efficiencies also are among the highest reported for solid-state LECs based on CTMCs. Such results indicate that the host–guest system is essential and useful for achieving highly efficient solid-state LECs.

2.2 Color Tuning

Early developments of CTMCs for LECs were focused on ruthenium-based complexes, which typically emit in the red and orange–red part of the spectrum due to low metal-to-ligand charge transfer (MLCT) energies. Ester-substituted ligands have been found to red-shift the emission of [Ru(bpy)$_3$](PF$_6$)$_2$ [12, 13]. Ng et al. [41] published a series of polyimides containing a Ru complex on the main chain. The device showed deep-red EL centered at about 650 nm and near infrared (IR) emission centered near 750–800 nm. However, the efficiency of such device is rather low (0.1%). Bolink et al. [42] also have achieved a deep red emission, using bis-chelated Ru complex [Ru(tpy)(tpy-CO$_2$Et)](PF6)$_2$, where tpy is terpyridine. The ester substituents were found to significantly improve the device efficiency over the unsubstituted terpyridyl complex, but the device efficiency is still very low. Low efficiency is a general phenomenon of terpyridyl Ru complexes since the luminescent ^{3}MLCT state is quenched by low-lying ^{3}MC levels due to poor bite angles of the tpy ligands [43].

Iridium was shown to have promise for color tuning over ruthenium complexes because of increased ligand-field splitting energies. The pioneering work of Ir-based complex ([Ir(ppy)$_2$(dtb-bpy)]PF$_6$ for LECs was reported by Slinker et al. [18]. Subsequent color tuning of CTMCs has centered on appropriate selection of ligand or transition metal core. Complexes based on fluorine substituents on the phenylpyridine ligands were reported to cause a blue shift of the emission spectrum relative to unfluorinated complexes. For example, ([Ir(ppy)$_2$(dtb-bpy)]PF$_6$ and [Ir(dF(CF$_3$)ppy)$_2$(dtb-bpy)]PF$_6$ showed EL centered at 542 nm and 500 nm, respectively [20, 21]. The increasing of bandgap of complexes results from the fact that metal-centered highest occupied molecular orbital (HOMO) level is highly stabilized by introduction of the fluorine atoms [21]. We also studied the electrochemical characteristics of complexes **1** and **2** via cyclic voltammetry. As shown in Fig. 9, complex **2** exhibits a reversible oxidation peak at 1.70 V vs. Ag/AgCl, which is significantly higher than that of complex **1** (1.35 V). The HOMO level of ([Ir(ppy)$_2$(dtb-bpy)]PF$_6$ had been reported to be a mixture of the d-orbitals of iridium and the π-orbitals of the phenyl ring of the ppy ligand [21]. Since **1** and **2** are also cationic ppy-based Ir(III) complexes, their HOMO distributions should be similar to that of ([Ir(ppy)$_2$(dtb-bpy)]PF$_6$. The higher oxidation potential of complex **2** compared with that of complex **1** indicates that the two electron-withdrawing fluoro substituents on the phenyl ring of the ppy ligand possibly reduce the electron density on the Ir metal center and consequently stabilize the HOMO level. According to the previous studies, the lowest unoccupied molecular orbital (LUMO) of ([Ir(ppy)$_2$(dtb-bpy)]PF$_6$ is located mainly on the dtb-bpy ligand [21], and therefore, for complexes **1** and **2**, reduction is expected to take place on the auxiliary SB ligand. Similar reduction potentials for complex **1** (–1.33 V) and complex **2** (–1.26 V) confirm that this common ligand is responsible for reduction of both compounds. Still, the reduction potential of complex **2** is slightly less negative than

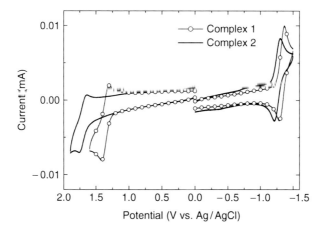

Fig. 9 Cyclic voltammograms of complexes **1** and **2**. Potentials were recorded vs. the reference electrode Ag/AgCl (sat'd). Reproduced with permission from [29]. Copyright 2007, WILEY-VCH Verlag GmbH & Co. KGaA, Weinheim

that of complex 1 perhaps because the fluoro substituents make complex 2 more electrophilic and thus easier to reduce [21]. With the HOMO level being significantly shifted and the LUMO level being only slightly changed by fluoro substitution, the PL emission of complex 2 is significantly blue shifted compared with PL of complex 1, reflecting a characteristic of Ir complexes that their energy gaps can be tuned through introducing substituents of different electronic properties on the ppy ligands.

Tamayo et al. [22] demonstrated tuning of the emission color across a large portion of the visible part of the spectra by independent tuning of the HOMO and LUMO levels of the parent complex [Ir(ppy)$_2$(bpy)]PF$_6$, where ppz is 1-phenylpyrazolyl. Highly efficient red (λ_{max} = 635 nm), green (λ_{max} = 542 nm), and blue (λ_{max} = 492 nm) EL emissions were obtained in LECs based on [Ir(tbppz)$_2$(biq)]PF$_6$, [Ir(ppy)$_2$(bpy)]PF$_6$, and [Ir(dF-ppz)$_2$(dtb-bpy)](PF$_6$), where tb is 5'-tert-butyl and biq is 2,2'-biquinoline [22].

Color tuning can also be achieved by destabilization of the LUMO rather than stabilization of the HOMO. Nazeeruddin et al. [44] achieved blue-green EL emission (λ_{max} = 520 nm) in LECs based on [Ir(ppy)$_2$(dma-bpy)]PF$_6$, where dma-bpy is 4,4'-(dimethylamino)-2,2'-bipyridine. Introduction of the electron-donating dimethylamino groups was demonstrated to significantly destabilize the LUMO relative to the HOMO when compared with the parent complex [Ir(ppy)$_2$(dtb-bpy)]PF$_6$, resulting in blue-shifted emission.

Attaching a charged side group on the periphery of the cyclometalating ligands proposed by Bolink et al. [45] converted a neutral complex into a CTMC. A blue-green emission (λ_{max} = 487 nm) was observed for [Ir(ppy-PBu$_3$)3](PF$_6$)$_3$. Such a blue shift in emission, compared with the [Ir(ppy)$_3$] complex, was attributed to the electron-withdrawing nature of the tri-butylphosphonium group on the phenylpyridine ligand. However, EL was found to change from blue-green to yellow (λ_{max} = 570 nm) after 100 s of operation at 4 V.

Recently, we have also developed a novel blue-green emitting ppz-based Ir-complex (stable EL λ_{max} = 488 nm) and a saturated red emitting ppy-based Ir-complex (PL λ_{max} = 672 nm). Details of photophysical and EL properties of these two complexes will be described in Sect. 3.

2.3 Lengthening Device Lifetime

The lifetimes of the devices defined as the time it takes for the brightness of the device to decay from the maximum to half of the maximum under a constant bias are important figure-of-merit when evaluating the durability of devices for display or lighting applications. Much effort has been made to lengthen device lifetimes of LECs.

The rates of electrochemical reactions of emissive CTMCs under electrical driving are bias dependent. In general, running LEC devices at a higher luminance leads to shorter lifetimes, so long lifetime can often be claimed at the expense of luminance [12, 29, 46–49]. The lifetimes for complex **1** under the 2.6-V driving and for complex **2** under the 2.9-V driving are ~26 h (extrapolated) and 6.7 h, respectively (Fig. 4). Although operated under a lower current density, the lifetime of the complex **2** device is significantly shorter than that of the complex **1** device. Irreversible multiple oxidation and subsequent decomposition under a high electric field had been proposed as the mechanism for degradation of LECs based on cationic iridium complexes [32, 50]. Since a higher bias voltage is required for operating the device based on complex **2** due to its larger energy gap, the higher electric field in the emissive layer of the complex **2** device perhaps accelerated the degradation. To mitigate the device degradation, slightly lower bias voltages of 2.5 V and 2.8 V were then applied for testing the devices based on complex **1** and complex **2**, respectively. As shown in Figs. 4a and 4b, slight reduction of the applied bias by 0.1 V had led to greatly reduced degradation rates for brightnesses of both devices. The lifetimes of the complex **1** device under the 2.5-V driving and the complex **2** device under the 2.8-V driving are ~54 and ~12 h (both extrapolated), roughly two times longer than those driven at 0.1-V higher bias voltages. The longer device lifetimes, however, are achieved at the expense of reduced brightness. Therefore, emissive materials with high quantum yields are essential for LECs to be operated at a practical brightness yet using a lowest possible bias, which would ensure long-lifetime operation of LECs.

Blending CTMCs in an inert polymer can also improve lifetime [14, 16]. It may result from lowered current density caused by separation of conducting chromophores in LEC devices. The choice of metal contacts was found to influence the lifetime, even for devices stored in the off state because reactive metal (e.g., Al) may activate some electrochemical reactions with CTMCs and lead to degradation [14, 51]. Degradation of $[Ru(bpy)_3]^{2+}$-based device involves the water-induced formation of photoluminescence quenchers, which were suggested to be complexes such as $[Ru(bpy)_2(H_2O)_2]^{2+}$ and $[Ru(bpy)_2(H_2O)]_2O^{4+}$ [36, 52, 53]. Thus, CTMCs with phenanthroline ligands credited with improved hydrophobicity increased resistance toward water-induced substitution reactions, leading to the higher device stability. $[Ir(ppy)_2(dp\text{-}phen)]PF_6$ device, where dp-phen is 4,7-diphenyl-1,10-phenanthroline, was found by Bolink et al. to have a lifetime of 65 h, which is the highest reported for a neat-film Ir-based LEC to date [54].

2.4 Shortening Turn-On Time

Turn-on time is defined as the time required for LEC devices to reach maximum emission under dc bias. For practical applications, turn-on times must be significantly reduced; however, most schemes for improving turn-on time come at a cost to stability.

Higher applied biases can lead to faster turn-on, but results in shorter lifetimes [12, 48]. To achieve faster turn-on without losing the stability, a pulsed-biasing scheme – a short pulse at a higher voltage was used to turn the device on, followed by a lower voltage that was used to drive the device for extended periods, was proposed by Handy et al. [12]. However, such technique complicates the driving circuits and the integration for applications. Reducing the thickness of the emissive layer of LECs also can reduce the turn-on time, but leads to a decrease in the device efficiency due to exciton quenching near the electrodes [55]. Smaller counterions such as BF_4^- or ClO_4^- lead to faster turn-on times than larger one, e.g., PF_6^-, but also deteriorate device lifetimes [13, 16]. Increasing ionic conductivity of the emissive layer by addition of an electrolyte to the CTMC layer has been shown to improve device lifetimes due to additional ions [20, 32, 56, 57]. However, device lifetimes were deteriorated as well.

Chemically attaching ionically charged ligands to metal complexes represents an alternative and promising approach to increase the ionic conductivity and to improve the turn-on speed of LECs, since the phase compatibility is less an issue in such configurations. Bolink et al. reported a homoleptic iridium complex containing ionically charged ligands, giving a net charge of +3 for each complex [45]. The LEC based on such complex showed a much improved turn-on time, yet also suffered other issues, e.g., color shift during operation and low efficiencies, which are relevant to practical applications as well. Zysmen-Colman et al. recently also reported homoleptic ruthenium-based and heteroleptic iridium-based CTMCs containing tethered ionic tetraalkylammonium salts [58]. Similarly, LECs constructed using such complexes also exhibited reduced turn-on times; yet EL efficiencies from these complexes are not satisfactory, compared with general EL efficiencies achievable with CTMCs. Hence, further studies on CTMCs containing ionically charged ligands that can give improved turn-on times, stable operation, and also high device efficiency are still highly desired.

In one of our previous works, we demonstrated the reduction in turn-on times of single-component Ir-based LECs with tethered imidazolium moieties. The new CTMC is achieved by fusing two imidazolium groups at the ends of the two alkyl side chains of a more conventional complex $[Ir(ppy)_2(dC6\text{-}daf)]PF_6$ (**3**) (where dC6-daf is 9,9′-dihexyl-4,5-diazafluorene), forming the new complex [Ir(ppy)$_2$(dCMIM-daf)](PF$_6$)$_3$ (**4**) (where dC6MIM-daf is 9,9-bis[6-(3-methylimidazolium)hexyl]-1-yl-4,5-diazafluorene) [59]. Molecular structures of complexes **3** and **4** are shown in Fig. 10.

Homogeneous thin films of both complexes **3** and **4** can be obtained by spin-casting from CH_3CN solution. We also tried to prepare a physical blending film of **3** and ionic liquid [BMIM][PF$_6$] for comparative studies; however, the resulting cloudy films indicated strong phase separation and incompatibility between complex **3** with the long alkyl chains and the ionic liquid with high polar character. The UV-visible absorption and PL spectra of **3** and **4** in CH_2Cl_2 solution (10^{-5} M) and in neat film (~100 nm) are shown in Figs. 11a and 11b, respectively. Similar absorption features are observed in either solutions or films for both complexes. Solution PL emission is in the yellow range (~570 nm) for both complexes **3** and **4**, indicating tethering imidazolium groups at the ends of the alkyl chains in complex **3** has no significant effects on modulating energy gaps.

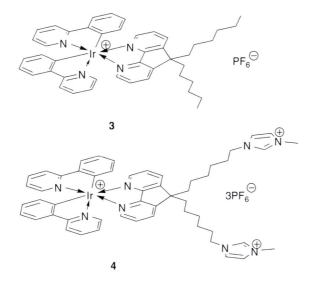

Fig. 10 Molecular structures of complexes **3** and **4**

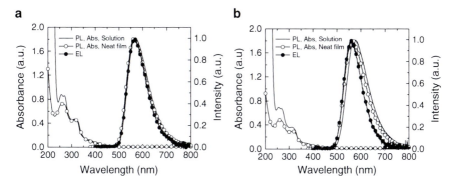

Fig. 11 Absorption and PL spectra in CH_2Cl_2 solution (10^{-5} M) and in neat film along with EL spectra under a forward bias voltage of 2.8 V for (**a**) complex **3** and (**b**) complex **4**, respectively. Reproduced with permission from [59]. Copyright 2008, WILEY-VCH Verlag GmbH & Co. KGaA, Weinheim

The PLQYs of complexes **3** and **4** in CH_2Cl_2, as determined with a calibrated integrating sphere, are 0.59 and 0.47, respectively (Table 3). Room-temperature transient PL of complexes **3** and **4** in CH_2Cl_2, as determined by the time-correlated single photon counting technique, exhibits the single-exponential decay behavior, and the extracted excited-state lifetimes for **3** and **4** are 1.12 μs and 0.94 μs, respectively (Table 3). Spin-coated neat films of complexes **3** and **4** exhibit PL spectra similar to those observed for their solutions (Fig. 11). The measured PLQYs and excited-state lifetimes of the neat films are (0.33, 0.66 μs) for **3** and (0.35, 0.75 μs) for **4** (Table 3). Lower PLQYs and shorter excited-state lifetimes in neat

Table 3 Summary of physical properties of complexes **3** and **4**

Complex	$\lambda_{max,\,PL}$ (nm)[a]		PLQY[b]		τ (μs)[c]		$E^{ox}_{1/2}$ (V)[d]	$E^{red}_{1/2}$ (V)[e]	$\Delta E_{1/2}$ (V)[f]
	Solution[g]	Neat film	Solution[g]	Neat film	Solution[g]	Neat film			
3	568	562	0.59	0.33	1.12	0.66	+1.33	−1.47	2.80
4	570	562	0.47	0.35	0.94	0.75	+1.31	−1.48	2.82

[a]PL peak wavelength
[b]Photoluminescence quantum yields
[c]Excited-state lifetimes
[d]Oxidation potential
[e]Reduction potential
[f]The electrochemical gap $\Delta E_{1/2}$ is the difference between $E^{ox}_{1/2}$ and $E^{red}_{1/2}$
[g]Measured in CH_2Cl_2 (10^{-5} M) at room temperature
Reproduced with permission from [59]. Copyright 2008, WILEY-VCH Verlag GmbH & Co. KGaA, Weinheim

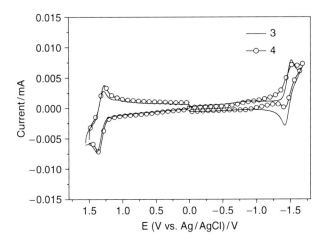

Fig. 12 Cyclic voltammogram of complexes **3** and **4**. All potentials were recorded vs. Ag/AgCl (sat'd) as a reference electrode. Reproduced with permission from [59]. Copyright 2008, WILEY-VCH Verlag GmbH & Co. KGaA, Weinheim

films indicate that interaction between closely packed molecules provides additional deactivation pathways. However, PLQYs of complexes **3** and **4** in neat films still retain ~56% and ~75% of those in solutions. Highly retained PLQYs in neat films of complexes **3** and **4** are likely to be associated with their sterically bulky dC6-daf and dC6MIM-daf ligands. It is noted that self-quenching in neat films of complex **4** is not as severe as that in neat films of complex **3** perhaps due to better steric hindrance provided by imidazolium groups in complex **4**.

Figure 12 depicts the electrochemical characteristics of complexes **3** and **4** probed by cyclic voltammetry and the measured redox potentials are summarized in Table 3. Complexes **3** and **4** exhibit reversible redox peaks at similar potential (vs. Ag/AgCl); one reversible oxidation potential occurs at 1.33 V and 1.31 V for **3** and **4**, respectively, while one reversible reduction potential occurs at −1.47 V and −1.48 V for **3** and **4**, respectively. The HOMO of [Ir(ppy)$_2$(N^N)]PF$_6$ complexes has been reported to be a mixture of the iridium and the phenyl group of the C^N ligand, while the LUMO mostly localizes on the N^N ligand [21]. Since the structures of complexes **3** and **4** differ only in the terminal alkyl chain of N^N ligand, their HOMO and LUMO distribution should be similar and give similar redox potentials. The presence of imidazolium moieties in **4** exhibited limited influence on the electrochemical properties of the Ir-center, especially the frontier molecular orbitals.

Device characteristics based on the structure of [glass substrate/ITO/complex **3** or **4** (100 nm)/Ag] are discussed and are summarized in Table 4. EL spectra of the devices based on complexes **3** and **4** are shown in Figs. 11a and 11b, respectively, for comparison with their PL spectra. EL spectra are basically similar to PL spectra, indicating similar emission mechanisms. Time-dependent brightness and current densities of the LEC devices based on complex **3** operated under 2.7 and 2.8 V are shown in Fig. 13a. The currents of devices increased slowly with time after the bias was applied and kept increasing even after 10-h operation. The turn-on time of the device ($t_{turn-on}$), defined as the time to achieve brightness of 1 cd cm^{-2}, was 65 min under 2.8 V. The brightness increased with the current and reached the maximum of 105 cd m^{-2} at ~500 min under 2.8 V. For devices under 2.7 V, the brightness reached 56 cd m^{-2} and was still increasing at 600 min. Comparatively, the LECs based on complex **4** showed much faster response. As shown in Fig. 13b, this device turned on sharply in 12 min under 2.8 V, it took only ~200 min for the devices to reach the maximal brightness of 79 cd m^{-2} and 31 cd m^{-2} under 2.8 V and 2.7 V, respectively. Such improved turn-on times confirm that the chemically tethering imidazolium groups onto complex **4** increase the density of mobile counterions (PF$_6^-$) in neat films and consequently increase the neat-film ionic conductivity.

Table 4 Summary of the LEC device characteristics based on complexes **3** and **4**

Complex	Bias (V)	$\lambda_{max,\ EL}$ (nm)[a]	$t_{turn-on}$ (min)[b]	t_{max} (min)[c]	L_{max} (cd m^{-2})[d]	$\eta_{ext,\ max},\ \eta_{p,\ max}$ (%, lm W^{-1})[e]
3	2.8	566	65	500	105	5.5, 18.7
	2.7		84	>600	>56[f]	5.5, 19.2
4	2.8	577	12	200	79	4.0, 14.7
	2.7		16	200	31	4.5, 17.1

[a]EL peak wavelength
[b]Time required reaching the brightness of 1 cd cm^{-2}
[c]Time required to reach the maximal brightness
[d]Maximal brightness achieved at a constant bias voltage
[e]Maximal external quantum efficiency and maximal power efficiency achieved at a constant bias voltage
[f]Maximal brightness was not reached during a 10-h measurement
Reproduced with permission from [59]. Copyright 2008, WILEY-VCH Verlag GmbH & Co. KGaA, Weinheim

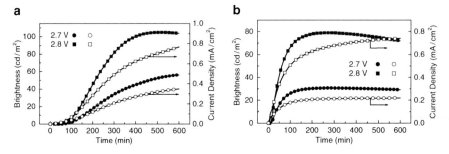

Fig. 13 Time dependent brightness and current density under 2.7 V and 2.8 V for the single-layered LEC device based on (**a**) complex **3** and (**b**) complex **4**, respectively. Reproduced with permission from [59]. Copyright 2008, WILEY-VCH Verlag GmbH & Co. KGaA, Weinheim

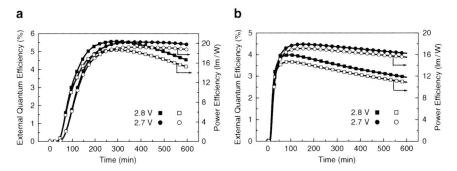

Fig. 14 Time-dependent EQE and the corresponding power efficiency under 2.7 V and 2.8 V for the single-layered LEC device based on (**a**) complex **3** and (**b**) complex **4**, respectively. Reproduced with permission from [59]. Copyright 2008, WILEY-VCH Verlag GmbH & Co. KGaA, Weinheim

Efficient LECs are essential for operation at a practical brightness yet using a lowest possible bias, which would ensure stable operation of LECs. Hence, it is important to examine the effect of chemical tethering of imidazolium groups with an ionic iridium complex on the device efficiencies of LECs. Time-dependent EQEs and corresponding power efficiencies of the complex **3** and **4** devices under biases of 2.7 V and 2.8 V are shown in Figs. 14a and 14b, respectively. When a forward bias was just applied, the EQE was rather low due to unbalanced carrier injection. During the formation of the p- and n-type regions near electrodes, the balance of the carrier injection was improved and the EQE of the device thus increased rapidly. The peak EQE and the peak power efficiency are (5.5%, 18.7 lm W^{-1}) and (5.5%, 19.2 lm W^{-1}) for the complex **3** device under biases of 2.8 V and 2.7 V, respectively. The time-dependent evolution trend in device efficiency of the complex **4** device is similar to that of the complex **3** device, yet it took much less time (within 100 min) for the complex **4** device to reach the peak device efficiency. The peak EQE and the peak power efficiency are (4.0%,

14.7 lm W^{-1}) and (4.5%, 17.1 lm W^{-1}) for the complex **4** device under 2.8 V and 2.7 V, respectively. EL efficiencies of devices based on both complexes **3** and **4** are comparable, indicating chemical tethering of imidazolium groups with an ionic iridium complex benefits faster device response yet does not influence EL efficiencies significantly. Hence, the technique of integrating ionic liquid moieties onto the CTMC directly serves as a useful and promising alternative to improve the turn-on speed of LECs and to realize single-component CTMC LECs compatible with simple driving schemes.

3 White Light-Emitting Electrochemical Cells Based on Cationic Transition Metal Complexes

Although solid-state LECs possess several advantages attractive for lighting applications, rare demonstration of white-emitting solid-state LECs had been reported. The only previous report of solid-state white LECs was based on phase-separated mixture of a polyfluorene derivative and polyethylene oxide (PEO) [10]. However, the fluorescent nature of conjugated polymers would limit the eventual EL efficiency. Recently, CTMCs have also been used in solid-state LECs [11–22, 31, 32, 36, 38, 41, 42, 44–49, 51–60], which show two major advantages over conventional polymer LECs (a) no ion-conducting material (e.g., PEO) is needed since these metal complexes are intrinsically ionic; and (b) higher EL efficiencies could be achieved due to the phosphorescent nature of the transition metal complexes. Inspired by previous works regarding energy transfer between ionic complexes reported by Malliaras [38] and De Cola [60], we had previously demonstrated highly efficient solid-state LEC devices by adopting host–guest cationic metal complexes [49]. Yet to our knowledge, there is no white LECs based on CTMCs ever being reported to date, despite their high potential.

Efficient white light emission may be most easily achieved by mixing two complementary colors, such as blue-green and red emission. Thus, the development of efficient blue-green [21, 22, 44, 60] and red-emitting [11–17, 22, 31, 38, 42, 48, 56] CTMCs is highly desired. In this section, we report the characterization of efficient blue-green and red-emitting cationic iridium complexes and their successful application in white LECs with adopting the effective host–guest strategy [38, 49, 61].

3.1 Photophysical and Electrochemical Properties of Novel Blue-Green and Red-Emitting Cationic Iridium Complexes

Molecular structures of complexes **5–8** are shown in Fig. 15. The emission properties of complexes **5–8** in solutions (DCM, 10^{-5} M) or in thin films are summarized in Table 5. For reducing the response time of LECs, in this work the ionic liquid

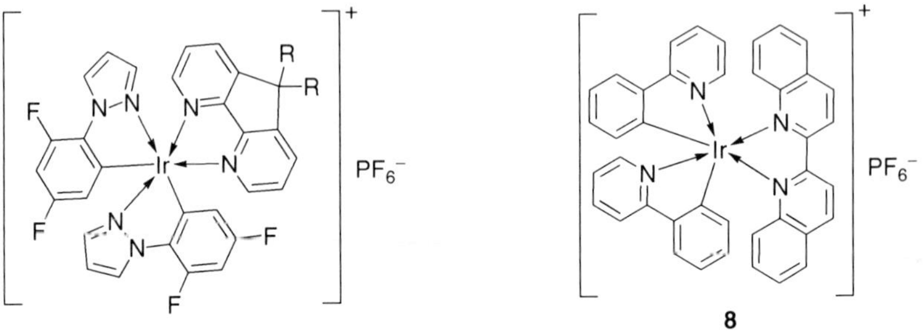

5: [Ir(dfppz)$_2$(dasb)]$^+$ (PF$_6^-$), R, R = 2, 2'-biphenyl
6: [Ir(dfppz)$_2$(bmpdaf)]$^+$ (PF$_6^-$), R, R = 4-CH$_3$OPh, 4-CH$_3$OPh
7: [Ir(dfppz)$_2$(dedaf)]$^+$(PF$_6^-$), R, R = C$_2$H$_5$, C$_2$H$_5$

Fig. 15 Molecular structures of complexes **5–8**

[BMIM][PF$_6$] was introduced in the emissive layer to provide additional mobile anions [29, 32, 49]. Thus, emission properties of **5–7** films in the presence of [BMIM][PF$_6$] (19 wt%) were also examined. In general, complexes **5–7** show PL peaks at 491–499 nm and rather high PLQYs of 0.46–0.66 in solutions. High PLQYs of these complexes indicate that the rigidity of the daf ligand is beneficial for reducing nonradiative (e.g., vibration) deactivation processes. Among the three blue-green complexes **5–7**, **7** exhibits shortest emission wavelengths of 488–491 nm and highest PLQYs of 0.28–0.30 in thin films (neat or dispersed with [BMIM][PF$_6$], and thus was chosen as the host material for white LEC studies. The absorption and PL spectra of blue-green complex **7** and red-emitting complex **8** in solutions and in neat films (~100 nm) are explicitly shown in Fig. 16. Interestingly, complex **7** with the alkyl substitution on daf exhibits nearly same emission wavelengths in solutions and in solid films. Complex **8** exhibits more saturated red emission (with PL peaks of 656 and 672 nm in solutions and in neat films, respectively) compared with that of [Ir(ppyz)$_2$(biq)]PF$_6$ [22]. The energy gaps estimated by cyclic voltammetry for complex **8** (2.23 eV, Table 5) and [Ir(ppyz)$_2$(biq)]PF$_6$ (2.45 eV) [22] are consistent with the photophysical observation. For [Ir(ppyz)$_2$(biq)]PF$_6$, the HOMO and the LUMO are predominantly localized on the ppz and biq ligand, respectively [22]. The smaller energy gap of complex **8**, which also contains biq, is thus associated with destabilization of HOMO in replacing ppz with ppy.

Cyclic voltammetry was used to probe the electrochemical properties of these complexes. The results are summarized in Table 5. For complexes **5–7**, one reversible oxidation potential was detected, which can be attributed to the oxidation that occurred on the Ir center. Interestingly, the electronic nature of ancillary daf ligands plays a subtle role on the oxidation potential. For example, complex **7** containing a more electro-rich dedaf ligand shows a lower oxidation potential (1.20 V) when compared with those of complexes **5** (1.29 V) and **6** (1.28 V), both

Table 5 Summary of physical properties of complexes 5–8

Complex	$\lambda_{max, PL}$ (nm), Φ, τ (μs)[a]			$E^{ox}_{1/2}$ (V)[b]	$E^{red}_{1/2}$ (V)[b]	$\Delta E_{1/2}$ (V)
	Solution[c]	Neat film or host–guest film	Film with [BMIM][PF$_6$][d]			
5	499, 0.46, 0.71	513, 0.20, 0.36	504, 0.22, 0.42	+1.29[e]	−1.70[b]	2.99
6	497, 0.66, 0.85	507, 0.22, 0.51	498, 0.26, 0.56	+1.28[e]	−1.72[f]	3.00
7	491, 0.54, 0.73	491, 0.28, 0.55	488, 0.30, 0.74	+1.20[g]	−1.79[f]	2.99
8	656, 0.20, 0.75	672, 0.09, 0.33	–	+0.86[f]	−1.37[f]	2.23
8 (0.2 wt %) : 7	–	(493, 606), 0.26, (0.49[h], 1.54[i])	(488, 603), 0.29, (0.54[h], 1.68[i])	–	–	–
8 (0.4 wt %) : 7	–	(493, 614), 0.27, (0.38[h], 1.52[i])	(488, 612), 0.28, (0.47[h], 1.68[i])	–	–	–

[a]At room temperature
[b]Potential vs. ferrocene/ferrocenium redox couple
[c]Measured in CH$_2$Cl$_2$ (10^{-5} M)
[d]Films containing 19 wt% [BMIM][PF$_6$]
[e]0.1 M tetra-n-butylammonium hexafluorophosphate (TBAPF$_6$) in acetonitrile
[f]0.1 M TBAP$_6$ in acetonitrile
[g]0.1 M TBAPF$_6$ in CH$_2$Cl$_2$
[h]Measured at 480 nm
[i]Measured at 650 nm
Reproduced with permission from [61]. Copyright 2008, American Chemical Society

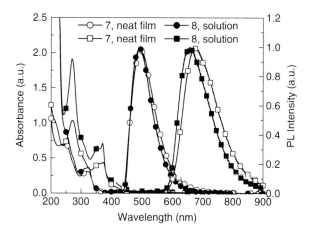

Fig. 16 Absorption (*left*) and PL spectra (*right*) of complexes **7** and **8** in dichloromethane solutions (10^{-5} M) and in neat films. Reproduced with permission from [61]. Copyright 2008, American Chemical Society

containing aryl substitutions on C9 of the diazafluorene ligand. On the contrary, the reduction capability of the diazafluorene ligand is presumably responsible for observed reversible reduction of complexes 5–7. Thus, the Ir complex coordinated to an electron-rich dedaf ligand (complex **7**) shows a more negative reduction potential (−1.79 V) when compared with those complexed to the aryl-substituted daf ligand [complexes **5** (−1.70 V) and **6** (−1.72 V)]. These results thus suggest that the energies of frontier orbitals in these complexes can be subtly manipulated by tailoring the electronic properties of the auxiliary daf ligands. For the red complex **8**, one reversible oxidation (0.86 V) and one reversible reduction (−1.37 V) were detected. The electron-rich character of the ppy ligand and the more π-delocalized biq ligand contribute to the higher ease of oxidation and reduction, respectively, thus leading to a small energy gap of complex **8**.

Emission properties of the **7**:**8** host–guest films are also summarized in Table 5. Figure 17 depicts the PL spectra of the host–guest films with different guest **8** concentrations (0.4 and 0.2 wt%) with or without the presence of [BMIM][PF$_6$] (19 wt%). In general, with the presence of [BMIM][PF$_6$], both the host and guest exhibit longer excited-state lifetimes and the PLQYs are slightly raised (Table 5), indicating the role of [BMIM][PF$_6$] in suppressing intermolecular interactions [49]. As shown in Fig. 17, white emission of different blue-green and red compositions can be achieved by adjusting the guest concentration. Raising the guest concentration effectively enhances the host–guest energy transfer, accompanied by shorter lifetimes of the host emission (Table 5). The host–guest film containing 0.4 wt% of **8** (guest) shows white emission having CIE coordinates of $(x, y) = (0.34, 0.37)$, which is rather close to the ideal equal-energy white $(x, y) = (0.33, 0.33)$, and therefore was subjected to further EL studies.

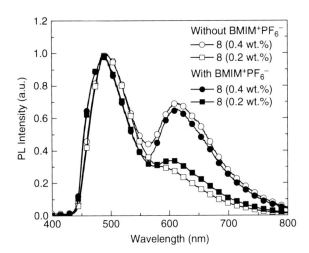

Fig. 17 PL spectra of host–guest films containing different guest concentrations (0.4 and 0.2 wt%) without and with [BMIM][PF$_6$] (19 wt%). Reproduced with permission from [61]. Copyright 2008, American Chemical Society

3.2 Electroluminescent Properties of White Light-Emitting Electrochemical Cells Based on Host–Guest Cationic Iridium Complexes

The LECs have the structure of ITO/emissive layer (100 nm)/Ag (150 nm), where the emissive layer contains host **7** (80.5 wt%), guest **8** (0.4 wt%), and [BMIM][PF$_6$] (19.1 wt%). Their EL properties are summarized in Table 6. The EL spectra of the white LECs under various biases, along with the PL spectra for comparison, are shown in Fig. 18a. It is noted that the peak wavelength of the blue component in EL spectra is 488 nm, which is one of the shortest EL wavelengths reported to date for LECs based on CTMCs. Compared with PL, the relative intensity of the red emission with respect to the blue emission is larger in EL and increases as the bias decreases. Furthermore, the maximum current density of the host–guest device (with 0.4 wt%

Table 6 Summary of white LEC device characteristics

Bias (V)	CIE (x, y)[a]	CRI[a]	t_{max}[b] (min)	L_{max}[c] (cd m^{-2})	$\eta_{ext, max}, \eta_{L, max}, \eta_{p, max}$[d] (%, cd A^{-1}, lm W^{-1})	$t_{1/2}$[e] (h)
2.9	(0.45, 0.40)	81	240	2.5	4.0, 7.2, 7.8	8.9
3.1	(0.37, 0.39)	80	60	18	3.4, 6.1, 6.2	1.3
3.3	(0.35, 0.39)	80	30	43	3.3, 5.8, 5.5	0.4

[a]Evaluated from the EL spectra
[b]Time required to reach the maximal brightness
[c]Maximal brightness achieved at a constant bias voltage
[d]Maximal external quantum efficiency, current and power efficiencies achieved at a constant bias voltage
[e]The time for the brightness of the device to decay from the maximum to half of the maximum under a constant bias voltage
Reproduced with permission from [61]. Copyright 2008, American Chemical Society

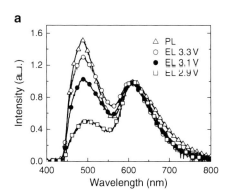
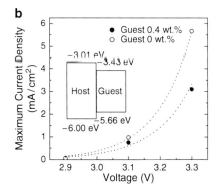

Fig. 18 (a) The bias-dependent EL spectra of the white LECs compared with the PL spectrum. (b) Maximum current density vs. voltage characteristics for the host–guest (0.4 wt% doping concentration) and the host-only LECs. *Inset*: the energy level diagram of the host and guest molecules. Reproduced with permission from [61]. Copyright 2008, American Chemical Society

guest) is lower than that of the host-only device under same bias conditions (Fig. 18b). These results could be understood by energy level alignments of the host and guest (inset of Fig. 18b). At lower biases, such energy level alignments favor carrier injection and trapping on the smaller-gap guest, resulting in direct carrier recombination/exciton formation on the guest (rather than host– guest energy transfer). Therefore, larger fractions of guest emission are observed at lower biases. Nevertheless, white emission with CIE coordinates of (0.35, 0.39) could be achieved at higher biases and brightness. Also, the present white LECs exhibit a rather high color rendering index (CRI) (up to 80), which is an important characteristic for solid-state lighting.

Figure 19a shows the time-dependent brightness and current density under constant biases of 2.9–3.3 V for the white LEC. After the bias was applied, the current first rose and then stayed rather constant. On the contrary, the brightness first increased with the current and reached the maxima of 2.5 cd m^{-2}, 18 cd m^{-2}, and 43 cd m^{-2} under biases of 2.9 V, 3.1 V, and 3.3 V, respectively. The brightness then dropped with time with a rate depending on the bias voltage (or current). Corresponding time-dependent EQEs and power efficiencies of the same device are shown in Fig. 19b. When a forward bias was just applied, the EQE was rather low due to poor carrier injection. During the formation of the doped regions near electrodes, the capability of carrier injection was improved and the EQE thus rose rapidly. The peak EQEs, current and power efficiencies at 2.9 V, 3.1 V, and 3.3 V are (4.0%, 7.2 cd A^{-1}, 7.8 lm W^{-1}), (3.4%, 6.1 cd A^{-1}, 6.2 lm W^{-1}), and (3.3%, 5.8 cd A^{-1}, 5.5 lm W^{-1}), respectively.

Peak brightness and turn-on time (the time required to reach the maximal brightness) as a function of bias voltage for white LECs are shown in Fig. 20a. An electrochemical junction between p- and n-type doped layers of LECs is formed during device operation. As revealed in previous studies [31], as bias voltage increases, the junction width decreases due to extension of doped layers, consequently leading to higher current density (Fig. 19a) and higher brightness (Fig. 20a). The higher electric field in the device also accelerates redistribution of mobile ions,

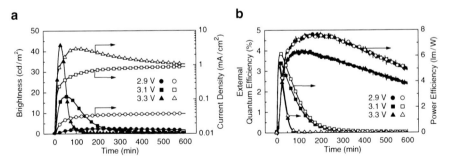

Fig. 19 (a) Brightness (*solid symbols*) and current density (*open symbols*) and (b) external quantum efficiency (*solid symbols*) and power efficiency (*open symbols*) as a function of time under a constant bias voltage of 2.9–3.3 V for the white LEC. Reproduced with permission from [61]. Copyright 2008, American Chemical Society

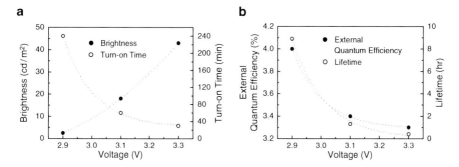

Fig. 20 (a) Peak brightness (*solid symbols*) and turn-on time (*open symbols*) and (b) peak external quantum efficiency (*solid symbols*) and lifetime (*open symbols*) as a function of bias voltage for the white LECs. Reproduced with permission from [61], American Chemical Society

which facilitates formation of ohmic contacts with the electrodes. Thus, operation of LECs under higher bias voltages speeds up the device response (Fig. 20a). However, higher brightness and faster response are obtained at the expense of device stability. As shown in Fig. 20b, peak EQE and device lifetime (the time for the brightness of the device decaying from the maximum to half of the maximum under a constant bias voltage) of the white LECs deteriorate under higher bias voltages. It may be associated with the fact that the higher electric field or current density accelerates degradation (multiple oxidation and subsequent decomposition) [32] of the emissive CTMCs. Previous studies showed that addition of materials altering the electronic properties of the emissive layer of LECs also influence the device lifetimes [14, 56]. Blending an inert polymer [14] in the emissive layer of LECs improves device lifetimes, while addition of an electrolyte [56] shortens device lifetimes. Blending of an inert polymer reduces the electronic conductivity of the emissive layer and thus lowers the current density. On the contrary, additional mobile ions provided by the electrolyte increase the doping level of the doped region, resulting in higher current density due to increased electronic conductivity. These results also imply that higher current leads to shorter device lifetimes. Recently, two reports [52, 53] have revealed that electrical driving of LECs based on CTMCs induces oxo bridged dimers, which effectively quench EL. However, detailed degradation mechanisms of the LECs based on CTMCs remain unclear and further studies are still needed to achieve practical device lifetimes.

4 Outlook

Solid-state CTMC-based LECs exhibit advantages of simple fabrication processes, low-voltage operation, and high power efficiency, which thus may be competitive for lighting applications. However, improvements will be required in our prototype white CTMC-based LECs for practical applications. For device efficiencies, measured device efficiency did not reach the upper limit (~6%) estimated from the PLQY of the emissive layer (~30%) and ~20% light out-coupling efficiency from a typical layered light-emitting device structure. It may be associated with exciton quenching by one of the electrodes if there is discrepancy in electron and hole transport, rendering the recombination zone closer to one of the electrodes. To further improve the device efficiency, further tailoring of molecular structures to achieve more balanced electron and hole transport is required. Improving device lifetimes will also be essential for lighting applications. Introducing functional groups with resistance against quencher formation and/or electrochemical degradation in CTMCs will facilitate realization of long-lifetime white LECs for lighting applications.

References

1. Kido J, Hongawa K, Okuyama K et al (1994) Appl Phys Lett 64:815
2. Tokito S, Iijima T, Tsuzuki T et al (2003) Appl Phys Lett 83:2459
3. D'Andrade BW, Forrest SR (2004) Adv Mater 16:1585
4. Gong X, Wang S, Moses D et al (2005) Adv Mater 17:2053
5. Huang J, Li G, Wu E et al (2006) Adv Mater 18:114
6. Tang CW, VanSlyke SA (1987) Appl Phys Lett 51:913
7. Burroughes JH, Bradley DDC, Brown AR et al (1990) Nature 347:539
8. Pei Q, Yu G, Zhang C et al (1995) Science 269:1086
9. Pei Q, Yang Y, Yu G et al (1996) J Am Chem Soc 118:3922
10. Yang Y, Pei Q (1997) J Appl Phys 81:3294
11. Lee JK, Yoo DS, Handy ES et al (1996) Appl Phys Lett 69:1686
12. Handy ES, Pal AJ, Rubner MF (1999) J Am Chem Soc 121:3525
13. Gao FG, Bard AJ (2000) J Am Chem Soc 122:7426
14. Rudmann H, Rubner MF (2001) J Appl Phys 90:4338
15. Liu CY, Bard AJ (2002) J Am Chem Soc 124:4190
16. Rudmann H, Shimada S, Rubner MF (2002) J Am Chem Soc 124:4918
17. Liu CY, Bard AJ (2005) Appl Phys Lett 87:061110
18. Slinker JD, Gorodetsky AA, Lowry MS et al (2004) J Am Chem Soc 126:2763
19. Lowry MS, Hudson WR, Pascal RA et al (2004) J Am Chem Soc 126:14129
20. Slinker JD, Koh CY, Malliaras GG et al (2005) Appl Phys Lett 86:173506
21. Lowry MS, Goldsmith JI, Slinker JD et al (2005) Chem Mater 17:5712
22. Tamayo AB, Garon S, Sajoto T et al (2005) Inorg Chem 44:8723
23. Wu CC, Lin YT, Chiang HH et al (2002) Appl Phys Lett 81:577
24. Wong KT, Chien YY, Chen RT et al (2002) J Am Chem Soc 124:11576
25. Wu CC, Liu TL, Hung WY et al (2003) J Am Chem Soc 125:3710
26. Wu CC, Lin YT, Wong KT et al (2004) Adv Mater 16:61
27. Chao TC, Lin YT, Yang CY et al (2005) Adv Mater 17:992
28. Wong KT, Chen RT, Fang FC et al (2005) Org Lett 7:1979
29. Su HC, Fang FC, Hwu TY et al (2007) Adv Funct Mater 17:1019
30. Kawamura Y, Goushi K, Brooks J et al (2005) Appl Phys Lett 86:071104
31. Rudmann H, Shimada S, Rubner MF (2003) J Appl Phys 94:115
32. Parker ST, Slinker JD, Lowry MS et al (2005) Chem Mater 17:3187
33. Yang C, Sun Q, Qiao J et al (2003) J Phys Chem B 107:12981

34. Edman L, Pauchard M, Moses D et al (2004) J Appl Phys 95:4357
35. Wenzl FP, Pachler P, Suess C et al (2004) Adv Funct Mater 14:441
36. Kalyuzhny G, Buda M, McNeill J et al (2003) J Am Chem Soc 125:6272
37. Tang CW, VanSlyke SA, Chen CH (1989) Appl Phys Lett 65:3610
38. Hosseini AR, Koh CY, Slinker JD et al (2005) Chem Mater 17:6114
39. Chen FC, Yang Y, Pei Q (2002) Appl Phys Lett 81:4278
40. Sinanoğlu O (1965) Modern quantum chemistry. Academic, New York
41. Ng WY, Gong X, Chan WK (1999) Chem Mater 11:1165
42. Bolink HJ, Cappelli L, Coronado E et al (2005) Inorg Chem 44:5966
43. Sauvage JP, Collin JP, Chambron JC et al (1994) Chem Rev 94:993
44. Nazeeruddin MK, Wegh RT, Zhou Z et al (2006) Inorg Chem 45:9245
45. Bolink HJ, Cappelli L, Coronado E et al (2006) Chem Mater 18:2778
46. Maness KM, Terrill RH, Meyer TJ et al (1996) J Am Chem Soc 118:10609
47. Maness KM, Masui H, Wightman RM et al (1997) J Am Chem Soc 119:3987
48. Bernhard S, Barron JA, Houston PL et al (2002) J Am Chem Soc 124:13624
49. Su HC, Wu CC, Fang FC et al (2006) Appl Phys Lett 89:261118
50. Ohsawa Y, Sprouse S, King KA et al (1987) J Phys Chem 91:1047
51. Rudmann H, Shimada S, Rubner MF et al (2002) J Appl Phys 92:1576
52. Soltzberg LJ, Slinker JD, Flores-Torres S et al (2006) J Am Chem Soc 128:7761
53. Slinker JD, Kim JS, Flores-Torres S et al (2007) J Mater Chem 17:76
54. Bolink HJ, Cappelli L, Coronado E et al (2006) J Am Chem Soc 128:14786
55. Lee KW, Slinker J, Gorodetsky AA et al (2003) Phys Chem Chem Phys 5:2706
56. Lyons CH, Abbas ED, Lee JK et al (1998) J Am Chem Soc 120:12100
57. Leprêtre JC, Deronzier A, Stephan O (2002) Synth Met 131:175
58. Zysman-Colman E, Slinker JD, Parker JB et al (2008) Chem Mater 20:388
59. Su HC, Chen HF, Wu CC et al (2008) Chem Asian J 3:1922
60. Coppo P, Duati M, Kozhevnikov VN et al (2005) Angew Chem Int Ed 44:1806
61. Su HC, Chen HF, Fang FC et al (2008) J Am Chem Soc 130:3413

Horizontal Molecular Orientation in Vacuum-Deposited Organic Amorphous Films

Daisuke Yokoyama and Chihaya Adachi

Contents

1 Introduction 138
2 Experiments and Calculations 140
   2.1 Variable Angle Spectroscopic Ellipsometry 140
   2.2 Cutoff Emission Measurement 142
3 Results and Discussion 144
4 Summary 150
References 150

Abstract Organic amorphous films fabricated by vacuum deposition have been used as essential components in organic light-emitting diodes (OLEDs) because they have the advantages of nanometer-scale surface smoothness and easy controllability of thickness. However, molecular orientation in organic amorphous films has been disregarded for the past 20 years since the beginning of the research on OLEDs. Here, we demonstrate horizontal molecular orientation in neat and doped organic amorphous films and show the general relationship between molecular structures and the molecular orientation. It was found that when molecular structure is linear or planar, the anisotropy of the molecular orientation in films becomes generally large. The results show the vital importance of the horizontal molecular orientation to understand the light emission and the carrier transport in OLEDs. To elucidate the device physics in OLEDs, the molecular orientation in amorphous films should generally be taken into consideration.

D. Yokoyama (✉)
Graduate School of Science and Engineering, Yamagata University, 4-3-16 Johnan, Yonezawa, Yamagata 992-8510, Japan
e-mail: d_yokoyama@yz.yamagata-u.ac.jp

C. Adachi (✉)
Center for Organic Photonics and Electronics Research (OPERA) / Center for Future Chemistry, Kyushu University, 744 Motooka, Nishi, Fukuoka 819-0395, Japan
e-mail: adachi@cstf.kyushu-u.ac.jp

1 Introduction

Organic amorphous films, fabricated by vacuum deposition, have played an important role in the development of organic light-emitting diodes (OLEDs) [1] and organic laser devices [2, 3]. Amorphous films, which have smooth interfaces, are essential for the fabrication of fine structures in devices that control light and charge carriers, whereas randomly rough interfaces, due to polycrystalline textures, for example, cause undesirable light scattering or leak current. The amorphous films also greatly contribute to the manufacture of practical, pinhole-free multilayer films using simplified fabrication techniques. These have, in turn, significantly increased the rate of development of organic thin film devices for the past 20 years since the beginning of the research on OLEDs [1].

However, it has been assumed that the full optical and electrical potentials of organic molecules cannot be achieved because it has been taken for granted that the molecules in organic vacuum-deposited amorphous films are randomly oriented and the films themselves are isotropic. In OLEDs fabricated by vacuum deposition, for example, light out-coupling efficiency is limited by the randomly oriented transition dipole moments of the emitting molecules [4–6]. If we can make the orientation of the transition dipole moments of the emitting molecules in OLEDs completely parallel to the substrate, the light out-coupling efficiency becomes ~50% higher compared to when the orientation of the transition dipole moments is random [6]. In organic semiconductor laser devices, the random orientation is also unfavorable because the alignment of the transition dipole moments of the emitting molecules makes it easy for stimulated emissions to occur. Furthermore, the alignment of molecules can facilitate charge carrier transport because it increases the overlap of π-orbitals between adjacent molecules.

Although complete alignment of molecules occurs naturally in organic single crystals, it is difficult to prepare single crystal thin films having a desired thickness and a smooth flat surface on a nanometer scale using a simple fabrication technique. Thus, instead of preparing ideal molecular single crystal films, the correct strategy is to produce a quasimolecular orientation in a smooth organic amorphous film in order to fabricate large-area, high-performance organic devices with low cost. Here, we pay particular attention to this viewpoint and investigate the molecular orientation in neat and doped organic amorphous films and its mechanism.

The molecular orientation in thin films can be detected by measuring the optical anisotropy in the films. One of the best methods of investigating the optical properties of thin films is by spectroscopic ellipsometry [7]. In particular, variable angle spectroscopic ellipsometry (VASE) is very sensitive to optical anisotropy in films and has been widely used to determine their optical constants (Fig. 1). VASE has also been used to investigate many kinds of anisotropic films made from spin-coated polymers [8–11] and is a reliable technique for determining the optical properties of both inorganic and organic films. Organic amorphous films are also good subjects for VASE measurements because it is easier to analyzing thin films

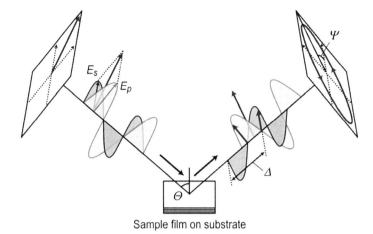

Fig. 1 Schematic illustrations of experimental setups of variable angle spectroscopic ellipsometry (VASE). Linearly polarized light is incident to an organic film deposited on a Si substrate with different incident angles of Θ. The ellipsometry parameters Ψ and Δ, which represent the ratio of amplitudes of s- and p-polarized components of the incident light and the phase difference between them, respectively, are obtained for multiple incident angles and wavelengths

having a smooth surface and subsequently determine their optical constants. Recently, Lin et al. [12] first reported the application of this technique to vacuum-deposited organic amorphous films and they detected uniaxial anisotropies in ter (9,9-diarylfluorene)s thin films. They also demonstrated that the molecular orientation affects the amplified spontaneous emission threshold of the films [13].

The second method, used to detect anisotropy in films, is the measurement of emissions at cutoff wavelengths [14] from the edges of substrates. It can only be applied to light-emitting films with smooth surfaces (Fig. 2) [15–21]. With this method, a characteristic spectrum having a narrow band at the cutoff wavelength can be observed in the direction parallel to the substrate surface. As the polarization characteristics of the emissions are related to the direction of the transition dipole moments of the emitting molecules, optical anisotropy in films can be detected. By this cutoff emission measurement (CEM) [15–20], the anisotropy of spin coated polymers has been investigated. However, a detailed analysis of the relationship between the peak wavelengths of the emissions and the anisotropic refractive indices has not yet been completed.

Recently, we observed spectrally narrow emissions from organic vacuum-deposited amorphous films [21]. Here, we demonstrate that they originate in the cutoff phenomenon and that there are large anisotropies in vacuum-deposited amorphous films of fluorescent bis-styrylbenzene derivatives. Additionally, we found that the ordinary refractive indices of the films of the bis-styrylbenzene derivatives are significantly higher than the extraordinary refractive indices, showing a horizontal orientation of the molecules in the amorphous films fabricated by vacuum deposition [22]. We also showed the dependence of the anisotropy on the molecular structures and underlying layers and discussed the mechanism causing the

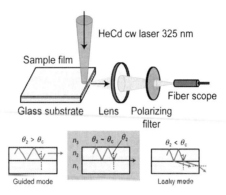

Fig. 2 Schematic illustrations of experimental setups of cutoff emission measurement (CEM). Organic films on glass substrates were optically pumped near the edges of the substrates by ultraviolet light from a cw He–Cd laser. Spectral shapes of edge emissions often significantly differ from the normal PL spectral shapes, and the peak wavelengths of the narrowed bands are in good agreement with the cutoff wavelengths of the slab waveguides that are composed of the glass substrate (with a refractive index of n_1), organic film (n_2), and air (n_3). The cutoff emission occurs as a boundary phenomenon between the cases where light is guided and where light is leaky, which means that the internal incident angle in the organic film θ_2 corresponds to the critical angle of the waveguide θ_c

anisotropy in vacuum-deposited amorphous films. Furthermore, we demonstrated the optical anisotropy in the films of the bis-stylylbenzene derivatives by using CEM in addition to VASE. A detailed analysis into the relationship between the peak wavelengths of the cutoff emissions and the anisotropic refractive indices was also performed. Although the two methods, VASE and CEM, are completely independent of each other, as seen in Figs. 1 and 2, respectively, there was an excellent agreement between their results, showing the reliability of the results obtained by the two different methods. Furthermore, by using CEM on doped amorphous films, we also found that the doped molecules adopted horizontal orientations even in isotropic host matrix films.

We discovered that molecules in many amorphous layers of hole transport materials (HTMs) and electron transport materials (ETMs) are horizontally oriented [23]. We report on the molecular orientation in films of various HTMs and ETMs and demonstrate that the molecular orientation is closely related to the driving voltage in OLEDs. Using wide-range VASE, we investigated the anisotropy of the molecules in the films of the HTMs and ETMs.

2 Experiments and Calculations

2.1 Variable Angle Spectroscopic Ellipsometry

Figure 3 shows all the materials used for the VASE analysis. All the samples of organic films used for the ellipsometry measurements were prepared by vacuum

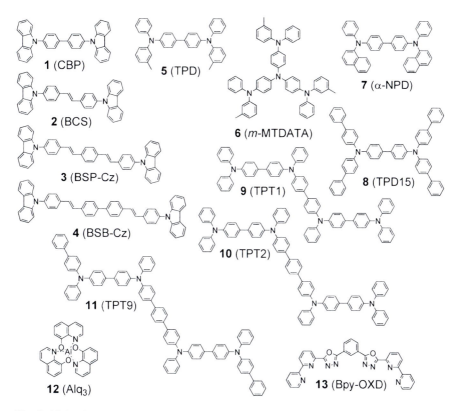

Fig. 3 Molecular structures of materials used in this study

deposition. We selected styrylbenzene derivatives (materials **1–4**) having carbazole groups at both ends of the molecules. Although it seems that, from the molecular formulas, these styrylbenzene derivatives have planar structures, they are actually twisted to avoid steric repulsion. We have also previously demonstrated that bis-styrylbenzene derivatives are laser-active materials, whose doped films have extremely low ASE thresholds [24]. The neat films fabricated by vacuum deposition are amorphous. We also selected HTMs (materials **5–11**) and ETMs (**12** and **13**) to investigate the relationship between the molecular structures and the optical anisotropies. Some of them are well-known materials used in OLEDs.

For materials **1–4**, the sample films were deposited on silicon(100) substrates with a deposition rate of 0.2 nm s^{-1}, and for materials **5–13**, the sample films were deposited on fused silica substrates with a deposition rate of 0.3 nm s^{-1}. All samples had a thickness of ~100 nm.

VASE was performed using a fast spectroscopic ellipsometer (M-2000U, J.A. Woollam Co., Inc.). Measurements were taken at seven multiple angles of the incident light from 45° to 75° in steps of 5°. At each angle, experimental ellipsometric parameters Ψ and Δ were simultaneously obtained in steps of 1.6 nm throughout the spectral region from 245 to 1,000 nm. It took only ca. 90 s to perform all of the measurements on a single sample. The analysis of the data was performed using the software "WVASE32" (J. A. Woollam Co., Inc.), which can perform a analysis including all data at all the different incident angles and wavelengths. The model function system of dielectric constants ε_1 and ε_2 of the organic films, which satisfies the Kramers–Kronig consistency, was composed of Gaussian and Tauc–Lorentz oscillators [25] with a Sellmeier background. Physically reasonable values were adopted as the initial values for the structural and optical parameters of the model functions. Then, the parameters were optimized to fit the simulated values of Ψ and Δ to the experimental values to minimize the value of the mean square error (MSE) [26]. The complex refractive indices of the films $N = n + ik$ were also calculated using the values determined for ε_1 and ε_2. An analysis using uniaxial anisotropic models [7] was also performed to determine the anisotropy of the films. Only the amplitudes of the oscillators were independently changed in the model function systems of the ordinary and extraordinary dielectric constants. The other parameters were changed with keeping the common values both for the ordinary and extraordinary constants in these systems.

2.2 Cutoff Emission Measurement

The organic films of tris(8-hydroxyquinoline)aluminum (Alq$_3$, material **12**), 4,4'-bis(N-carbazole)-biphenyl (CBP, material **1**), and 4,4'-bis[(N-carbazole)styryl]biphenyl (BSB-Cz, material **4**) were also prepared by vacuum deposition to measure cutoff emissions. The organic films, which had 15 different thicknesses, were deposited on clean glass substrates having a refractive index of 1.524. The glass substrates with the deposited organic films were then cleaved, and the organic films were optically pumped near to the edge of the cleavage face using a cw He–Cd laser (IK3102R-G, Kimmon Electric Co.). An incident beam with a wavelength of 325 nm was focused onto the surface of the organic films using a lens. The excitation energy of the beam was 60 W cm^{-2}. The edge emissions were collected using a multichannel spectrometer (PMA-11, Hamamatsu Photonics Co.) that had an optical fiber with a receiving diameter of 1 mm. The polarization characteristics of the edge emissions were measured through a polarizing filter placed between the sample and the end of the fiber.

The cutoff wavelengths of the waveguides with isotropic and uniaxial anisotropic organic core layers were calculated as below (also see Fig. 2). The wavelength of the light that propagates in a waveguide should satisfy the phase matching condition represented by:

$$\frac{4\pi n_2 d \cos\theta_2}{\lambda} + \phi_{21} + \phi_{23} = 2m\pi, \tag{1}$$

where d is the thickness of the core layer, λ is the free space wavelength of light, ϕ_{21} and ϕ_{23} are the phase shifts at the interfaces 21 and 23, respectively, and m is the mode number ($m = 0, 1, 2, \ldots$). The phase shifts depend on the polarization direction and are represented for the transverse electric (TE) and transverse magnetic (TM) modes by:

$$\phi_{2i}^{TE} = -2\tan^{-1}\frac{(n_2^2 \sin^2 \theta_2 - n_i^2)^{1/2}}{n_2 \cos \theta_2} \quad (i = 1 \text{ or } 3) \tag{2}$$

and

$$\phi_{2i}^{TM} = -2\tan^{-1}\frac{n_2(n_2^2 \sin^2 \theta_2 - n_i^2)^{1/2}}{n_i^2 \cos \theta_2} \quad (i = 1 \text{ or } 3) \tag{3}$$

respectively. If the wavelength of light is very close to the cutoff wavelength, the incident angle θ_2 corresponds to the critical angle θ_c, that is,

$$\theta_2 = \theta_c = \sin^{-1}(n_1/n_2). \tag{4}$$

From (1)–(4), we obtain the cutoff wavelength for the TE and TM modes,

$$\lambda_c^{TE} = \frac{2\pi d(n_2^2 - n_1^2)^{1/2}}{\tan^{-1}\left(\frac{n_1^2 - n_3^2}{n_2^2 - n_1^2}\right)^{1/2} + m\pi} \tag{5}$$

and

$$\lambda_c^{TM} = \frac{2\pi d(n_2^2 - n_1^2)^{1/2}}{\tan^{-1}\left[\frac{n_2^2}{n_3^2}\left(\frac{n_1^2 - n_3^2}{n_2^2 - n_1^2}\right)^{1/2}\right] + m\pi} \tag{6}$$

respectively. Here, d is the thickness of the organic film, and n_1, n_2, and n_3 are the refractive indices of the glass substrate ($n_1 = 1.524$), the organic film (n_2), and air ($n_3 = 1$), respectively. The refractive indices of the organic films (n_2) in all of the above equations depend on the wavelength of light. Using (5) and (6), we can directly calculate the cutoff wavelengths of a waveguide with an isotropic core layer at a certain thickness. When a core layer has a uniaxial anisotropic property, the calculations become more complicated because the refractive indices for the light of the TE and TM modes are different. The refractive index for the light of the TE mode n_2^{TE} is simply equal to the ordinary refractive index, while that of the TM mode n_2^{TM} depends on the angle θ_2, according to the equation for an index ellipsoid. They are represented by:

$$n_2^{TE} = n_o \tag{7}$$

and

$$\frac{1}{n_2^{TM}(\theta_2)^2} = \frac{\cos^2\theta_2}{n_o^2} + \frac{\sin^2\theta_2}{n_e^2} \tag{8}$$

respectively [27]. The cutoff wavelength for the TE mode is obtained by substituting (7) into (5). To calculate the cutoff wavelength for the TM mode, it is necessary to determine the angle θ_2 at each wavelength because the refractive index depends on this angle. The equation for the cutoff condition, (4), is replaced by

$$\theta_2 = \sin^{-1}\left(n_1/n_2^{TM}(\theta_2)\right). \tag{9}$$

From (8) and (9), we obtain the equation for determining θ_2,

$$n_1 - n_o n_e \left(n_o^2 \sin^2\theta_2 + n_e^2 \cos^2\theta_2\right)^{-1/2} \sin\theta_2 = 0. \tag{10}$$

Equation (10) can be solved for θ_2 numerically at each wavelength, and the n_2^{TM} determined by (8) is substituted into (6). In this way, we can calculate the cutoff wavelength in both cases: when the core layer is isotropic and when it is anisotropic.

3 Results and Discussion

Ellipsometric parameters of many films could not be correctly simulated using an isotropic model even if numerous oscillators were included in the analysis. However, this was not the case if a uniaxial anisotropic model was used. This shows the uniqueness of analysis by VASE. The results demonstrate the anisotropy of the films. We also checked whether the in-plane anisotropy exists in the films by rotating the sample, and this resulted in no in-plane difference.

The refractive indices and the extinction coefficients of all films obtained by VASE using a uniaxial model are shown in Fig. 4. Many films have uniaxial anisotropy and the ordinary refractive indices are much higher than the extraordinary ones. The ordinary extinction coefficients of the films are also much higher than the extraordinary ones.

In the case of BSB-Cz (material 4), the transition dipole moment of the BSB-Cz molecule is almost in line with the molecular long axis, and the ordinary extinction coefficient is higher than the extraordinary one. This means that the BSB-Cz molecules in the film lie nearly parallel to the substrate surface. The ordinary refractive index is also much higher than the extraordinary one. This is because π-electrons in the molecules can be moved by external electric fields more readily in the direction of the molecular long axis than in the direction perpendicular to it. Furthermore, no in-plane anisotropy means that the BSB-Cz molecules lie in random directions in the film. In our separate experiments, we could not observe apparent sharp peaks in the X-ray diffraction (XRD) patterns of the BSB-Cz film or

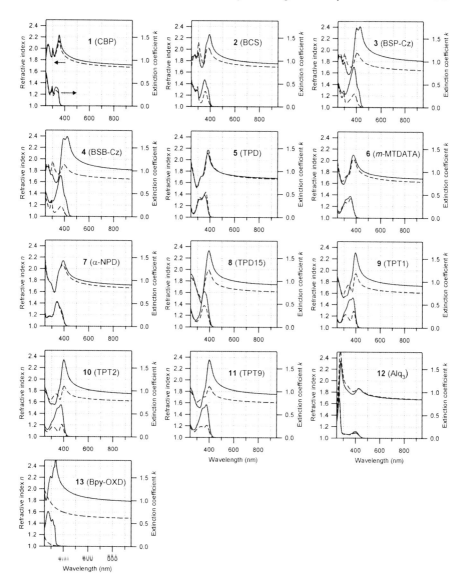

Fig. 4 Ordinary and extraordinary refractive indices (*left axis*) and extinction coefficients (*right axis*) of films. The ordinary and extraordinary components are indicated by *solid* and *broken lines*, respectively

any of the other films. From these results, we conclude that the BSB-Cz film is in a "horizontally oriented amorphous state," where molecules are horizontally oriented on the substrate, but there is no crystallographic structure with a long range order because the in-plane molecular orientation is random. Similar results for the horizontal molecular orientation in ter(9,9-diarylfluorene) thin films were reported by Lin et al. [12].

Interestingly, we found that the horizontal molecular orientation can occur on any underlying layers. The large anisotropies were observed in BSB-Cz films on a silicon substrate, a glass substrate, a 50-nm thick CBP layer, a 100-nm thick Ag layer, and a 30-nm thick smooth ITO layer with an RMS of 0.6 nm. These results are very important because they signify the high versatility of the horizontal orientation for various applications to organic devices, such as OLEDs, in which the emitting layer is usually sandwiched between electrodes and/or other organic layers.

Then, by comparing the anisotropy of the films of materials **1–4**, we found that a long molecular structure causes the large anisotropy. The longer the molecular length, the larger the birefringence Δn becomes. The birefringence of the films, however, does not directly reflect the anisotropy of the molecular orientation because each molecule itself has the anisotropy of molecular polarizability. To clarify the relationship between the molecular orientation and the molecular structure, we introduced an orientation order parameter S as an indicator of molecular orientation [28]; that is,

$$S = P_2(\cos\theta) = \frac{1}{2}\langle 3\cos^2\theta - 1 \rangle = \frac{k_e - k_o}{k_e + 2k_o}, \qquad (11)$$

where $P_2(x)$ is the second Legendre polynomial, $\langle \ldots \rangle$ indicates ensemble average, θ is the angle between the molecular long axis and the direction vertical to the substrate surface, and k_o and k_e are the ordinary and extraordinary extinction coefficients at the peak wavelength, respectively. $S = -0.5$ if the molecules are completely parallel to the surfaces, $S = 0$ if they are randomly oriented, and $S = 1$ if they are completely perpendicular to the surface. The last term in (11) can be calculated using extinction coefficients determined by VASE using uniaxial anisotropic models, which is based on the assumption that the transition dipole moment of the molecule is parallel to the molecular axis. The values of S for materials **1–4** are -0.07, -0.17, -0.29, and -0.33, respectively, demonstrating the relationship between the molecular orientation and the molecular length. This means the longer the molecular length, the larger the anisotropy of the molecular orientation becomes.

Then, focusing on HTMs (materials **5–11**), the ordinary refractive indices and extinction coefficients for materials **8–11** are much higher than the extraordinary ones, whereas those for materials **5–7** do not significantly differ. The higher ordinary refractive indices and extinction coefficients mean that there are horizontal molecular orientations in the amorphous films, as already explained in the case of material **4**. As the molecular lengths of materials **8–11** are high and the interaction between adjacent molecules is not very strong, the molecules lie on the substrate without aggregation. The large polarizability and transition dipole moment along the molecular axis cause the high ordinary refractive indices and extinction coefficients, respectively. It should be noted that the film of material **8** has a large anisotropy, although the shape of the molecule is planar rather than linear.

Table 1 MSE values converged using isotropic model (MSE$_{iso}$) and uniaxial anisotropic model (MSE$_{aniso}$) in ellipsometry analysis for films, glass transition temperature (T_g) [29], molecular weight (MW), orientation order parameter (S) of films of HTMs, and difference in driving voltage of OLEDs with thin and thick layers of HTMs (ΔV) [29]

Material	MSE$_{iso}$	MSE$_{aniso}$	T_g (°C) [29]	MW	S	ΔV (V) [29]
5 (TPD)	4.5	4.4	58	517	0.02	2.01
6 (m-MTDATA)	7.3	4.8	75	789	–0.04	1.90
7 (α-NPD)	8.6	6.5	96	589	–0.01	1.60
8 (TPD15)	17.9	8.7	132	793	–0.15	1.41
9 (TPT1)	16.5	8.3	144	975	–0.20	0.96
10 (TPT2)	15.8	3.8	150	1,051	–0.28	0.60
11 (TPT9)	17.4	7.1	155	1,204	–0.27	0.37

The relationship between the anisotropy and the electrical characteristics of the HTMs is also interesting. In Table 1, the molecular weight (MW) and orientation order parameters (S) are listed. We also list the difference in the driving voltages of OLEDs with thin (50 nm) and thick (300 nm) layers of HTMs (ΔV), which were reported previously [29]. A correlation between S and ΔV was obtained, suggesting that horizontally oriented molecules provide better electrical characteristics of films. The horizontal molecular orientation can create a large overlap between π-orbitals of adjacent molecules, leading to the facilitation of the carrier transport. Furthermore, although a correlation between T_g and ΔV was observed [29], it would indirectly reflect the effect of the molecular length and shape on both the thermal characteristics of the bulk samples and the molecular orientation in the films. Therefore, because a large MW and a long molecular structure lead to a high T_g [30] and a larger S, an indirect correlation between T_g and ΔV would be observed.

In the case of ETMs (materials **12** and **13**), a significantly large optical anisotropy could be seen for the film of material **13**, whereas the film of material **12** was isotropic. Ichikawa et al. [31, 32] reported that because material **13** has two specific stable geometries, its vacuum-deposited film forms an amorphous state without crystallization, although it does not have a bulky substituent. We think that because both stable geometries are planar and very thin, the molecules are well horizontally oriented, leading to high-performance electron transport characteristics and a significant difference between the ordinary and extraordinary refractive indices and extinction coefficients.

Next, we demonstrated that the molecular orientation based on the results of CEM. Figure 5a–d shows the cutoff emission spectra of the Alq$_3$, CBP, BSB-Cz, and 6 wt%-BSB-Cz-doped CBP films with 15 different thicknesses on glass substrates under optical excitation. We observed spectrally narrow bands that were strongly polarized in TE mode or TM mode. The electric field in the TE mode is polarized, parallel to the substrate surface, while that in the TM mode is polarized in the plane, perpendicular to it. All the TE and TM spectra in Fig. 5a–d are normalized such that the peak intensities of the TE spectra become unity, while retaining the ratios of the intensities of the TE and TM spectra. The ratios of the intensities of the TE and TM modes reflect both the direction of the transition dipole moment

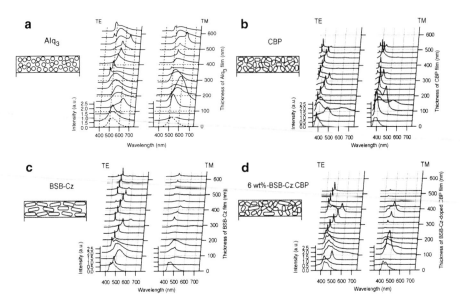

Fig. 5 Dependence of cutoff emission spectrum on polarization direction and thickness of organic film (**a**) Alq$_3$, (**b**) CBP, (**c**) BSB-Cz, and (**d**) 6 wt%-BSB-Cz-doped CBP. *Insets*: Schematic illustrations of molecular orientations in films. All the TE and TM emission spectra are normalized such that the peak intensities of the TE spectra become unity while keeping the ratios of intensities of the TE and TM spectra. Normal PL spectra are shown by *broken lines* in front

of the emitting molecules in the organic film and the intensity of the normal photoluminescent (PL) spectrum of the film at the peak wavelengths of the narrow TE and TM emissions. In the case of the isotropic Alq$_3$ and CBP films, the intensities of the TE and TM emissions were, overall, almost identical, as shown in Fig. 5a and 5b. In contrast, Fig. 5c shows that the intensities of the TE emission of the BSB-Cz films are much higher than that of the TM emissions. This result means that the transition dipole moments of the molecules in the film are nearly parallel to the polarization direction of the TE emissions. Thus, the horizontal orientation of the molecules on glass substrates is demonstrated by CEM, independently of VASE.

Furthermore, because the weak interaction between molecules causes the horizontal orientation, guest molecules having a linear-shaped structure can be horizontally oriented in host films. Figure 5d shows the cutoff emissions from the edges of 6 wt%-BSB-Cz-doped CBP films. It should be noted that the TM emissions from the BSB-Cz-doped CBP films were suppressed compared to the TM emissions from the neat CBP films shown in Fig. 5b. (Although the high TM intensities are apparently seen at the thicknesses of ~100 and ~380 nm in Fig. 5d, it does not mean random orientation or normal orientation. As the cutoff wavelengths of the TE modes at these thicknesses are accidentally out of the PL spectral region, as shown later, the absolute intensities of the TE emissions became low. Furthermore, because the cutoff wavelengths of the TM modes are inside the PL region, the intensities of the TM emissions became relatively higher compared to those of the TE mode at these thicknesses.) This result shows the horizontal orientation of the guest BSB-Cz molecules even in the isotropic host matrix of the CBP film. This can be well-understood if we assume that the surface of the films are always smooth on a nanometer-scale during codeposition and that the depositions of each BSB-Cz and CBP molecule are independent due to the weak interactions between them.

The validity of the results obtained by VASE and CEM are also shown in Fig. 6a and 6d, in which the peak wavelength of the cutoff emissions and the calculated cutoff wavelength of the films with different thicknesses are both shown. The former was obtained by CEM, and the latter was obtained by calculations using refractive indices obtained by VASE. They were obtained by the two independent methods using different substrates. The cutoff wavelengths of the slab waveguide, which is composed of a glass substrate, an isotropic organic film (Alq$_3$ or CBP), and air, were directly calculated using (5) and (6). As shown in Fig. 6a and 6b, the peak wavelength of the cutoff emissions and the calculated cutoff wavelength are in good agreement with each other. The peak wavelengths show a shift in the spectral region of the normal PL spectrum with an increase in the film thickness, and then the narrow

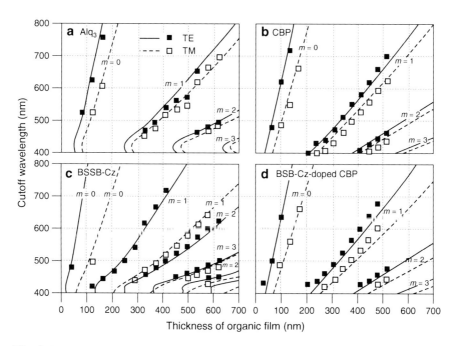

Fig. 6 Agreement between experimental peak wavelengths of cutoff emissions and calculated cutoff wavelengths of slab waveguides. The peak wavelengths of the TE and TM emissions (*open* and *closed squares*, respectively) and the cutoff wavelengths of the TE an TM emissions calculated using the refractive indices (*solid* and *broken lines*, respectively) are shown. (**a**) Alq$_3$, (**b**) CBP, (**c**) BSB-Cz, and (**d**) 6 wt%-BSB-Cz-doped CBP

emission having the next mode number appears from the short-wavelength side. The peak wavelengths of the TE and TM emissions are similar to each other, and the wavelengths of the TE emissions are slightly longer than those of the TM emissions, which are the characteristics of a slab waveguide with an isotropic layer.

In contrast, the peak wavelengths of the emission from the neat BSB-Cz films change in a different way, depending on the thickness, as shown in Fig. 6c. The peak wavelengths of the TE and TM emissions are not always similar. We succeeded in simulating this behavior by a calculation in which we took into account the anisotropy of the BSB-Cz film according to the process using the equation of the index ellipsoid, as shown in (8). The peak wavelengths and the calculated cutoff wavelength exactly matched with each other, demonstrating the validity of the results obtained by both VASE and CEM.

The peak wavelengths of the cutoff emissions from the BSB-Cz-doped CBP films almost corresponded to the calculated cutoff wavelengths of the waveguide with the CBP film, although slight deviations were found, as shown in Fig. 6d. The deviations in the TE modes in the short-wavelength region are due to perturbation by the doped BSB-Cz molecules because the ordinary refractive index of the BSB-Cz film is much higher than that of the CBP film.

4 Summary

In this study, we found that the linear- or planar-shaped molecules are horizontally oriented in organic amorphous films fabricated by conventional vacuum deposition. The general relationship between the molecular structure and the anisotropy of the orientation is clearly demonstrated. The weak interactions between adjacent molecules and linear or planar molecular structure cause the horizontal orientation. The fact that the horizontal molecular orientation occurs on any underlying layers, and even in doped film, shows the high versatility of the horizontal orientation for various applications. Our findings will lead to an improvement in optical and electrical characteristics of organic amorphous films and optoelectronic devices made from them. Moreover, there is a possibility that the horizontal molecular orientation is present in many amorphous films having high-performance optical or electrical characteristics. We think that it is meaningful to reevaluate the optical and electrical properties of organic materials in terms of their molecular orientation.

References

1. Tang CW, VanSlyke SA (1987) Appl Phys Lett 51:913–915
2. Kozlov VG, Bulovic V, Burrows PE et al (1998) J Appl Phys 84:4096–4108
3. Baldo MA, Holmes RJ, Forrest SR (2002) Phys Rev B 66:035321
4. Tsutsui T, Aminaka E, Lin CP et al (1997) Philos Trans R Soc Lond A 355:801–804

5. Gu G, Garbuzov DZ, Burrows PE et al (1997) Opt Lett 22:396–398
6. Kim JS, Ho PKH, Greenham NC (2000) J Appl Phys 88:1073–1081
7. Fujiwara H (2007) Spectroscopic ellipsometry: principles and applications. Wiley, New York
8. Miller EK, McGehee MD, Diaz-Garcia M et al (1999) Synth Met 102:1091–1092
9. Ramsdale CM, Greenham NC (2002) Adv Mater 14:212–215
10. Losurdo M, Giangregorio MM, Capezzuto P et al (2003) Synth Met 138:49–53
11. Winfield JM, Donley CL, Kim JS (2007) J Appl Phys 102:063505
12. Lin HW, Lin CL, Chang HH et al (2004) J Appl Phys 95:881–886
13. Lin HW, Lin CL, Wu CC et al (2007) Org Electron 8:189
14. Marcuse D (1991) Theory of dielectric optical waveguides, 2nd revised edn. Academic, San Diego
15. Peng X, Liu L, Wu J et al (2000) Opt Lett 25:314–316
16. Sheridan AK, Turnbull GA, Safonov AN et al (2000) Phys Rev B 62:R11929–R11932
17. Kawase T, Pinner DJ, Friend RH et al (2000) Synth Met 111–112:583–586
18. Pauchard M, Vehse M, Swensen J et al (2003) J Appl Phys 94:3543–3548
19. Li F, Solomesch O, Mackie PR et al (2006) J Appl Phys 99:013101
20. Yim KH, Friend RH, Kim JS (2006) J Chem Phys 124:184706
21. Nakanotani H, Adachi C, Watanabe S et al (2007) Appl Phys Lett 90:231109
22. Yokoyama D, Sakaguchi A, Suzuki M et al (2009) Org Electron 10:127–137
23. Yokoyama D, Sakaguchi A, Suzuki M et al (2008) Appl Phys Lett 93:173302
24. Aimono T, kawamura K, Goushi K et al (2005) Appl Phys Lett 86:071110
25. Synowicki RA, Tiwald TE (2004) Thin Solid Films 455–456:248–255
26. Jellison GE Jr (1993) Thin Solid Films 234:416–422
27. Saleh BEA, Teich MC (1991) Fundamentals of photonics. Wiley, New York
28. Ward IM (1975) Structure and properties of oriented polymers. Applied Science, London
29. Aonuma M, Oyamada T, Sasabe H et al (2007) Appl Phys Lett 90:183503
30. Fox TG Jr, Flory PJ (1950) J Appl Phys 21:581
31. Ichikawa M, Kawaguchi T, Kobayashi K et al (2006) J Mater Chem 16:221
32. Ichikawa M, Hiramatsu N, Yokoyama N et al (2007) Phys Stat Sol (RRL) 1:R37

Recent Advances in Sensitized Solar Cells

Arthur J. Frank

Contents

1 Introduction 154
   1.1 Disorder in Nanoporous Nanostructured TiO_2 Films 154
   1.2 Molecular Adsorbents 155
2 Experimental Methods 156
3 Oriented Nanotube Arrays 157
   3.1 Disordered Particle Films Versus Oriented Nanotube Films 157
   3.2 Orientational Disorder in Nanotube Films 159
4 Molecular Adsorbent Effects 165
5 Conclusions 167
References 168

Abstract Understanding the principal physical and chemical factors that govern or limit cell performance is critical for underpinning the development of next-generation sensitized solar cells. Recent studies of dye-sensitized solar cells (DSSCs) covering nanoporous (pore diameter <100 nm) one-dimensional TiO_2 nanostructured arrays and molecular voltage enhancers are discussed. Films constructed of oriented one-dimensional nanostructures, such as nanotube arrays, which are aligned perpendicularly to the charge-collecting substrate, could potentially improve the charge-collection efficiency by promoting faster transport and/or slower recombination. The extent to which transport or recombination could be affected by an oriented architecture is expected to depend on the influence of other mechanistic factors, such as the density and location of defects, crystallinity, and film uniformity. Orientational disorder within the nanotube array could also influence the transport and recombination kinetics. Such architectural disorder in titanium dioxide films is shown to have a strong influence on the transport, recombination, and light-harvesting properties of DSSCs. The mechanism by

A.J. Frank
Chemical and Materials Science Center, National Renewable Energy Laboratory, Golden, CO 80401, USA
e-mail: Arthur.Frank@nrel.gov

which molecular adsorbents alter the photovoltage of DSSCs is dependent on the properties of the adsorbent. In principle, an adsorbent could affect the photovoltage by either altering the recombination rate of photoelectrons in TiO_2 with oxidized redox species in the electrolyte or inducing band-edge movement. The net effect of altering the band positions and recombination kinetics can either improve or diminish cell performance. The mechanisms by which several molecular adsorbents increase the photovoltage of DSSCs are discussed.

1 Introduction

Sensitized mesoscopic nanostructured solar cells are considered one of the promising solar cell technologies of the future. The photoelectrode consists of sensitizers (e.g., dye molecules [1–5] or semiconductor quantum dots [6–8]) adsorbed or deposited onto a highly porous layer of a wide-bandgap nanocrystalline oxide, such as TiO_2. When photoexcited, the sensitizer injects electrons into the oxide support. The positive charge (hole) on the sensitizer is then transferred to a second phase – an ionic or electronic charge-conducting medium that fills the nanopores (pore diameter <100 nm) of the nanocrystalline layer. Electrons move through the nanostructured network and are collected at a transparent conducting oxide (TCO) substrate. The cycle is completed when the positive charges in the second phase are collected at the counter electrode. A critical feature of such cells is that light absorption and charge separation occur in close proximity – typically, on a length scale of <10 nm. Sensitized nanostructured solar cells include both hybrid organic/inorganic and entirely inorganic structures. The dye-sensitized solar cell (DSSC) is a prominent example of a general class of sensitized solar cells (SSCs). The traditional DSSC, which is often referred to as the Grätzel cell in recognition of the pioneering work of the inventor, utilizes a nanoporous (pore diameter <100 nm) nanoparticle titanium dioxide film covered with ruthenium bipyridyl-based dyes and triiodide/iodide as the redox relay in the electrolyte. The triiodide ions transport the holes from the oxidized dye molecules to the counter electrode. Certified sunlight-to-electricity conversion efficiencies of over 11% at full sunlight (AM1.5 solar irradiance) have been reported for such cells [9]. It is anticipated that by developing the scientific underpinning to exploit the unique properties of sensitized mesoscopic systems, efficiencies as high as 20% are attainable. Moreover, such studies may lay the scientific foundation for developing nanostructured systems with efficiencies beyond the Shockley–Queisser limit of 32% [10].

1.1 Disorder in Nanoporous Nanostructured TiO_2 Films

Mesoscopic nanocrystalline TiO_2 films used in traditional DSSCs have significant disorder associated with the individual particles (e.g., defects and size and shape nonuniformities) [11, 12], the particle network (e.g., distribution of the number of interparticle connections) [13, 14], and the interparticle contact area [15]. Electron transport in nanoporous nanoparticle TiO_2 films is 10^2–10^3 times slower than in single-crystal TiO_2 [16–18], a phenomenon attributable to some form of disorder associated with the presence of transport-limiting traps [13, 14, 19–25]. One of the important implications of traps limiting transport is that the longer the transport pathway, the more electrons undergo trapping and detrapping, and the more time they spend in a film before being collected. Thus, disorder associated with the particle network (versus the disorder associated with a single nanocrystal) is also expected to influence transport [13]. We have modeled the nanoporous TiO_2 films as a random particle network [14]. It was shown that the film porosity strongly influences the coordination number of the particles. For example, the average coordination number of particles ranges from about 5 at 50% porosity to about 2.5 at 75% porosity [13, 14]. Electron transport over such simulated nanoporous random nanoparticle TiO_2 films has been modeled using the random-walk approach [13]. Diffusion (random walk) of electrons through such simulated networks is described very well by percolation theory. It was estimated that for a typical film, increasing its porosity from 50 to 75% resulted in electrons visiting on average tenfold more particles, indicating that with higher film porosities, the electron transport pathway becomes more tortuous (longer) and electron transport becomes slower. Because the collection of photoinjected electrons competes with recombination, high charge-collection efficiency requires that transport be much faster than recombination. Thus, the film architecture is expected to strongly affect the electron transport and recombination dynamics and, correspondingly, the solar cell performance [13, 14].

One strategy to foster faster electron transport in DSSCs is to reduce both the morphological disorder and dimensionality of the network. The use of nanoporous films constructed of oriented one-dimensional nanotube (NT) [26–29] arrays, aligned orthogonally to the electron-collecting substrate, is representative of this approach. The use of such orientationally ordered NT films in DSSCs could potentially improve the charge-collection efficiency by promoting faster transport and/or slower recombination compared with the electron dynamics in randomly packed nanoparticle-based films. The extent to which transport or recombination could be affected by an oriented architecture is expected to depend on the importance of other mechanistic factors, such as the density and location of structural defects and the crystallinity. Film morphology is also expected to affect light harvesting properties of DSSCs [30, 31].

1.2 Molecular Adsorbents

Tailoring the interfacial properties is one strategy to improve cell performance. By simply exposing the nanocrystalline surface to certain molecules, the open-circuit photovoltage (V_{OC}) of a cell can be altered. Recombination of photoinjected electrons in TiO$_2$ films with oxidized redox species at the TiO$_2$/electrolyte interface represents a major loss of photoelectron density from the film and is, therefore, a major loss of photovoltage. Passivation of surface recombination sites is one mechanism by which an adsorbent can increase the photovoltage by reducing the loss of photoelectron density [32–35]. However, an adsorbent could also increase recombination, which would result in a lower V_{OC}. Band-edge movement is another mechanism by which an adsorbent can alter V_{OC} [36, 37]. Band-edge movement occurs when a sufficient net number of charges (or dipoles) build up on the surface of the particles to induce a change in the potential drop across the Helmholtz layer. A net negative surface charge buildup would cause the band edges to move upward, leading to a higher photovoltage. A net positive surface charge buildup would cause the band edges to move downward, resulting in a lower photovoltage.

In this chapter, I describe recent and interesting observations [29, 38] made mainly in the author's laboratory on the influence of morphological order in nanoporous TiO$_2$ films on the electron dynamics and light harvesting in DSSCs. The transport, recombination, and light-harvesting properties of dye-sensitized oriented NT arrays and randomly packed nanoparticle films are compared. The effects of orientational disorder within the NT array are explored. The mechanisms by which several molecular adsorbents affect the photovoltage of DSSCs are also discussed [37, 39].

2 Experimental Methods

Titanium dioxide NT arrays were prepared by electrochemically anodized Ti foils immersed in a fluoride-containing organic electrolyte in a two-electrode cell using a procedure discussed elsewhere [29]. The films were annealed in air at 400°C and afterward stained with the N719 dye (N719 = [tetrabutylammonium]$_2$[Ru(4-carboxylic acid-4′-carboxylate-2,2′-bipyridyl)$_2$(NCS)$_2$]), using the same procedure described for nanoparticle films [39]. The average particle size in a film was 24 nm. The amount of adsorbed N719 was measured by optical absorption of the desorbed dye as detailed previously [39]. The nanoparticle films had thicknesses ranging from 1 to 6 μm, a surface roughness factor of 90 μm^{-1}, and a porosity of 63%. The electrolyte used for the DSSCs was composed of 0.8 M 1-hexyl-2,3-dimethylimidazolium iodide and 50 mM iodine in methoxypropionitrile. The experimental conditions for employing the adsorbents tetrabutylammonium chenodeoxycholate (TBACDC) and guanidinium thiocyanate in DSSCs are described elsewhere [37, 39]. Scanning electron microscopy (SEM), transmission electron microscopy (TEM), and X-ray diffraction (XRD) were used to characterize the film morphologies. Transport and recombination properties were measured by intensity-modulated photocurrent spectroscopy (IMPS) and intensity-modulated photovoltage spectroscopy (IMVS), respectively.

3 Oriented Nanotube Arrays

3.1 Disordered Particle Films Versus Oriented Nanotube Films

Figure 1 displays SEM images of as-deposited NT films [29]. The NTs are packed in approximately hexagonal symmetry. The average NT had an inner pore diameter of about 30 nm and wall thickness of about 8 nm. The thicknesses of the NT films ranged from 1 to 6 μm depending on the anodization time. At the same film thickness, the NT films had about the same dye loading (98 ± 10%) as the nanoparticle films, implying that both films had comparable surface areas; their surface roughness factors were estimated to be about 90 μm^{-1} from gas desorption measurements of the nanoparticle films. The comparable roughness factors suggest that dye molecules cover both the interior and exterior walls of the NTs. From geometric considerations of the hexagonal arrangement of the NTs and knowledge of the roughness factor and wall thickness of the NT arrays, we estimate that the films had a porosity of 65% – similar to that of the nanoparticle films – and that the average center-to-center distance between NTs was about 56 nm, corresponding to an intertube spacing of 10 nm. The pore density of the NT films can be estimated from the pore diameter, wall thickness, and intertube spacing. The pore density of these films is quite high (ca 3.7×10^{10} pores cm^{-2}), owing to the close packing of the NTs. X-ray diffraction patterns of NT films (not shown) reveal that annealing in air transforms the as-deposited NT film from an amorphous material to anatase TiO_2. Sintering of the films had no effect on the pore diameter, wall thickness, and intertube spacing. The average crystallite size in the walls of the NTs, as calculated by application of the Scherrer equation [40] to the anatase (101) peak at $2\theta = 25.3°$,

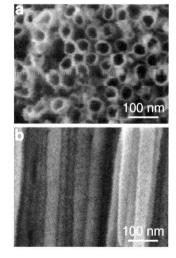

Fig. 1 SEM images of (**a**) surface and (**b**) cross-section of as-deposited TiO_2 NTs. Reprinted with permission from [29]. Copyright 2007, American Chemical Society

was 28 nm, although high-resolution TEM revealed the presence of crystallites longer than 50 nm.

Figure 2 shows the transport and recombination times of electrons in NT and nanoparticle films in DSSCs as a function of the incident light intensity (I_0) [29]. From Fig. 2a, it can be seen, perhaps counterintuitively, that the electron transport times for both film morphologies are comparable. The similar power-law dependence of transport on the light intensity for NT and nanoparticle films suggests that similar mechanistic factors govern transport. The power-law dependence of transport in nanoparticle films has been explained by a model in which electrons perform an exclusive random walk between trap sites that have a power-law distribution of waiting (release) times in the form of $t^{-1-\alpha}$, where the parameter α is related to disorder in the TiO$_2$ films [21, 41]. From the analyses of the power-law data in Fig. 2a using the expression [41]:

$$\tau_c \propto (I_0)^{\alpha-1}, \qquad (1)$$

where τ_c denotes the transport time constant, one can show that the amount of disorder associated with the trap density is larger in the NT films than in the nanoparticle films. The presence of a higher trap density in the NT films could explain the unexpectedly slow transport. On the other hand, Fig. 2b shows that recombination is ten times slower in the NT films than in the nanoparticle films. Given that recombination in the NT films is an order of magnitude slower than in the nanoparticle films, and that the transport times for both film morphologies are similar, the charge-collection efficiency (η_{cc}) for NT electrodes, as described by the relation [42] $\eta_{cc} = 1 - (\tau_c/\tau_r)$, is significantly higher, at least 25% larger in the NT-based DSSCs than in the nanoparticle-based counterparts at the maximum power point and at the highest light intensity (6.6×10^{16} cm^{-2}s^{-1}).

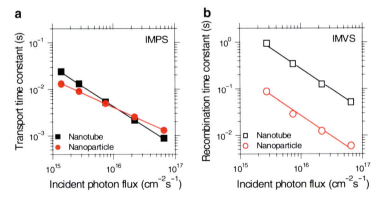

Fig. 2 Comparison of (**a**) transport and (**b**) recombination time constants for NT- and nanoparticle-based DSSCs as a function of the incident photon flux for 680-nm laser illumination. Reprinted with permission from [29]. Copyright 2007, American Chemical Society

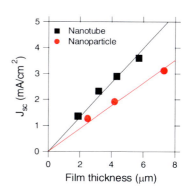

Fig. 3 Dependence of the short-circuit photocurrent densities on film thicknesses for dye-sensitized NT- and nanoparticle-based cells. The incident photon flux from a 680-nm laser was 6.6×10^{16} cm^{-2}s^{-1}cells. *Lines* represent linear fits of data. Reprinted with permission from [29]. Copyright 2007, American Chemical Society

Figure 3 shows the dependence of the short-circuit photocurrent density (J_{SC}) on the thickness of NT and nanoparticle films at 680-nm illumination [29]. For both film morphologies, the photocurrent densities increase linearly with film thickness. However, the slope of the plot for the NT films is 50% steeper than that of the nanoparticle films. The rate of increase of J_{SC} with film thickness is larger for the NT films than for the nanoparticle films. In general, J_{SC} can be approximated by the expression: $J_{SC} = q\eta_{lh}\eta_{inj}\eta_{cc}I_0$, where q is the elementary charge, η_{lh} is the light-harvesting efficiency of a cell, and η_{inj} is the charge-injection efficiency. The light-harvesting efficiency is determined by the amount of adsorbed dye, which is proportional to the film thickness for weakly absorbed light (as is the case for 680-nm illumination), the light-scattering properties of the film, the concentration of redox species, and other factors [43]; η_{inj} of adsorbed dye on the TiO$_2$ NTs and nanoparticles were assumed to be the same. From the analysis of this expression for J_{SC} and the observation that the charge-collection efficiency is 25% larger in the NT films, one can show that the light-harvesting efficiency of NT films is at least 20% greater than that of the nanoparticle films. Because both film morphologies have the same amount of dye, the higher light harvesting is attributed to increased light scattering in the NT films.

3.2 Orientational Disorder in Nanotube Films

An examination of the SEM image in Fig. 4 reveals that there is substantial architectural disorder in NT films. It can be seen that there are regions of clumps of NTs and crack-like features in the films. This observation is not limited to our NT films. There are numerous examples in literature showing the same architectural features in oriented TiO$_2$ NT-based films [44, 45]. Clusters of bundled NTs could be produced during the electrochemical synthesis of the as-deposited film [44, 46]. Alternatively, they could form during the washing and evaporative drying process of the as-deposited films via capillary forces of the liquid acting between the NTs.

Fig. 4 SEM image of TiO$_2$ NT film that was rinsed with water and then dried in air. Adapted with permission from [38]. Copyright 2007, American Chemical Society

Bundling due to capillary effects has been observed in materials that are relatively soft in the lateral direction, such as Si nanowires and carbon NTs. However, it is not self-evident that such bundling would occur in TiO$_2$ NTs because they are more rigid in the lateral direction. Whether produced during the formation of the NT arrays in the electrochemical synthesis or by capillary forces in the post-growth treatment of the as-deposited films, understanding the origin of such orientational disorder and its prevention is important for designing more ordered nanoporous film architectures.

The capillary force F between two adjacent nanotubes owing to a liquid meniscus can be determined from the equation: $F = \frac{1}{2}\pi\gamma(d_o cos\theta)^2 l^{-1}$, where γ is the surface tension of the liquid, θ is the contact angle, d_o is the external diameter of nanotube, and $2l$ is the center-to-center distance between NTs. The maximum lateral deflection (δ) of a NT resulting from capillary force exerted at the NT's end farthest from the substrate is given by the expression [38]:

$$\delta = \frac{32(d_o \cos\theta)^2(\gamma L^3)}{3El(d_o^4 - d_i^4)}. \qquad (2)$$

Equation (2) reveals that the maximum NT deflection strongly depends on surface tension value of the evaporated solvent, film thickness L, internal pore diameter d_i, wall thickness ($d_o - d_i$), Young's modulus, and the contact angle. For the case of a microcrack created by the formation of adjacent domains of bundled NTs with similar maximum lateral deflection, the maximum crack width (w_c) can be estimated with the expression: $w_c = 2\delta$.

Figure 5 shows SEM images of as-deposited NT films of different thicknesses after they were washed with water and then dried in air [38]. The surface tension value of water is about 72 dynes cm^{-1} (25°C), which is relatively high compared with organic liquids. The film thicknesses range from 1.1 to 6.1 μm. The presence of clusters of NT bundles and microcracks can be seen in even the thinnest sample – the 1.1-μm-thick film (Fig. 5a). When the film thickness increases from 1.1 to 6.1 μm (Fig. 5a–c), the extent of bundling and crack width increases consistent with (2), suggesting that the observed deformation is caused by the

Fig. 5 SEM images of (**a**) 1.1-, (**b**) 2.8-, and (**c**) 6.1-μm-thick as-deposited NT films that were rinsed with water and then dried in air. The *arrow* in image (**b**) marks a bundle of deflected NTs, and the *dashed line* signifies an associated crack. Reprinted with permission from [38]. Copyright 2007, American Chemical Society

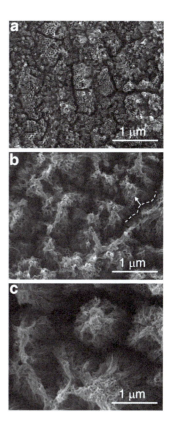

NT lateral deflection resulting from the action of capillary forces between adjacent NTs during evaporative drying of the wetted films. The dashed line (Fig. 5b) marks a boundary of a crack, and the arrow denotes a bundle of deflected NTs associated with it. The crack widths increase with film thickness from roughly 80 to 700 nm. In the thicker films, an overlayer covering clusters of bundled NTs can be seen.

Figure 6 displays SEM images of as-deposited NT films after they were washed with ethanol and air-dried; the surface tension value of ethanol is 22. Although there are still clusters of NT bundles and microcracks in the thinnest film, the extent of them is much less than those in the 1.1-μm-thick film (Fig. 5a) that was washed with water and air-dried. The extent of bundling and microcracks is also less in the ethanol-washed, air-dried thicker films compared with the corresponding thick films washed with water and air-dried.

Figure 7 shows SEM images of as-deposited NT films that were washed with ethanol and then dried using the supercritical CO_2 (scCO_2) technique [38]. (In its supercritical state, CO_2 passes from the liquid to gaseous phase without crossing the liquid–gas boundary and forming the associated interfacial tension (i.e., $\gamma = 0$ dynes cm^{-1}). In the absence of surface tension, there is no evidence of bundled

NTs, cracks, or overlayers in these three films with different thicknesses, in concurrence with the prediction from (2). These results are consistent with the hypothesis that capillary stress created during evaporation of liquids from the nanopores of as-deposited NT films can be of sufficient magnitude to cause bundling and microcrack formation. Furthermore, these results rule out the possibility that morphological disorder was produced during the electrochemical synthesis of the as-deposited NT films. Analyses of the SEM data in Figs. 5–7 show that the average deflection of the NT clusters increased with the film thickness and with the surface tension value of the evaporative liquid. However, the extent of deflection was quantitatively less than expected, suggesting that cooperative interactions of the neighboring NTs in a bundle of closely packed NTs limited the amount of bending movement of individual NTs.

From dye-desorption measurements, it was discovered that dye coverage was about 23% greater in the ethanol/scCO$_2$-dried films than in water-washed, air-dried films, suggesting that films with the more aligned NTs had a 23% larger surface area than films with substantial amounts of NT bundling. The smaller surface area of the more disordered architecture is attributed to the presence of NT bundling that

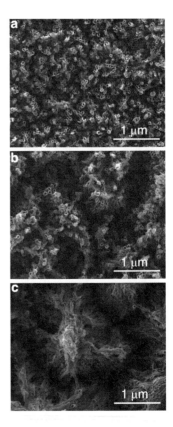

Fig. 6 SEM images of (**a**) 1.1-, (**b**) 2.8-, and (**c**) 6.1-μm-thick as-deposited NT films that were rinsed with ethanol and then dried in air. Reprinted with permission from [38]. Copyright 2007, American Chemical Society

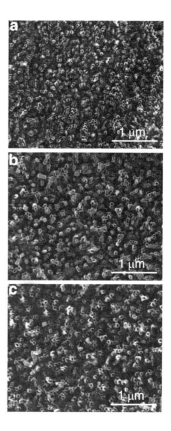

Fig. 7 SEM images of (a) 1.1-, (b) 2.8-, and (c) 6.1-μm-thick as-deposited NT films that were rinsed with ethanol and then dried using the scCO$_2$ technique. Reprinted with permission from [38]. Copyright 2007, American Chemical Society

blocks the accessibility of dye molecules to certain regions within a tightly packed NT bundle.

Figure 8 shows the effect of film morphology on the transport time constants as a function of light intensity for DSSCs [38]. The transport time constants for films that were water/air-dried or were ethanol/scCO$_2$-dried display the same power-law dependence on the light intensity, which means that they have the same α values (1), implying that the shapes of the distributions of localization (trapping) times are the same. From analyses of the transport time constants, one can deduce that transport is 75% faster in the ethanol/scCO$_2$-dried film than in the water/air-dried film. However, considering only the larger surface area of the ethanol/scCO$_2$-dried film, and therefore, the larger trap density, one might have expected [47] that transport would be 13% slower rather than 75% faster. If the contribution of the surface area to transport were removed by normalizing the surface areas of the films, one discovers that transport was actually two times faster. This would be understandable if electrons in a bundle of NTs could move either across the interconnecting NTs or along the length of a single NT toward the electron-collecting substrate. This situation would represent a mixture of one-dimensional

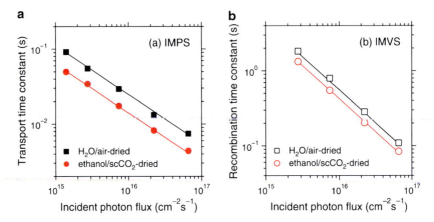

Fig. 8 Comparison of (**a**) transport and (**b**) recombination time constants as a function of the incident photon flux for 680-nm laser illumination for DSSCs, incorporating 6.1-μm-thick annealed NT films; preannealed (as-deposited) films were either rinsed with water and then air-dried or cleaned with ethanol and then scCO$_2$-dried. Reprinted with permission from [38]. Copyright 2007, American Chemical Society

(1D) and three-dimensional (3D) transport. Transport along 1D pathways should be three times faster than transport along pathways that are 3D. And so, the two-times-faster transport in the ordered 1D array is consistent with a transport mechanism of the disordered film being a mixture of 1D and 3D pathways. Thus, preventing disorder reduces the dimensionality of transport from a mixture of 1D and 3D transport to the expected 1D transport.

The time constant for recombination shows the same power-law dependence as does transport, which indicates a temporal connection between transport and recombination. From analyses of the time constants for recombination, one finds that recombination is 30% faster in the ethanol/scCO$_2$-dried films than in the water washed/air-dried films. Considering the larger surface area of the ethanol/scCO$_2$-dried films, one would have expected that recombination would be only 7% faster. After removing the contribution of surface area to recombination by normalizing the surface area, recombination is found to be only 22% faster. One would have predicted that if transport limited recombination, then recombination would be two times faster, the same as in the case of transport. This discrepancy between the observation and prediction is, in part, attributed to the creation of additional recombination centers (per unit surface area) in the water/air-dried film. It is envisioned that as the NTs bend to form bundles, the resulting distortion of the NTs increases the number of surface defects. These additional surface defect states (recombination centers) are present in the H$_2$O-washed/air-dried film, but are absent in the ethanol/scCO$_2$-dried film. Thus, recombination in the ethanol/scCO$_2$-dried film is significantly slower than the expected factor of two because of the absence of these additional surface defects. Thus, preventing NT bundling reduces the potential for additional traps that would have been otherwise created by the NT bending.

4 Molecular Adsorbent Effects

Chemical modification of the TiO_2 surface is a viable strategy for improving the performance of DSSCs. A simple approach to modify the interfacial properties is to expose the surface of TiO_2 films to certain molecular adsorbents in solution. In principle, such adsorbents can alter the band energetics and recombination kinetics in ways that can either increase or reduce cell efficiency. Here, we examine mechanistic effects of the adsorbents chenodeoxycholate and guanidinium (Fig. 9) on the photovoltage of dye-sensitized nanoparticle TiO_2 solar cells [37, 39].

Figure 10 shows the effect of the adsorbents chenodeoxycholate and guanidinium on V_{OC} over a range of photoinduced charge densities (n). V_{OC} varies logarithmically with n [36]:

$$V_{OC} = \frac{k_B T}{\alpha q} \ln\left(\frac{n}{n_0}\right), \qquad (3)$$

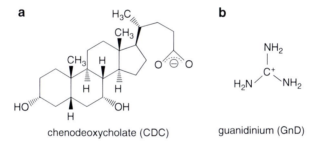

Fig. 9 Structures of adsorbents (**a**) chenodeoxycholate (CDC) and (**b**) guanidinium (GnD). Adapted with permission from [37] and [39]. Copyright 2005 and 2006, American Chemical Society

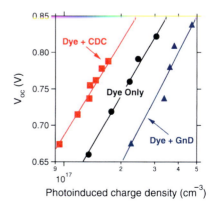

Fig. 10 Effect of the adsorbents chenodeoxycholate and guanidinium on the open-circuit photovoltages over a range of photoinduced charge densities at open circuit. Adapted with permission from [37] and [39]. Copyright 2005 and 2006, American Chemical Society

where q is the electronic charge, k_B is the Boltzmann constant, T is the absolute temperature, and n_0 is the electron density in the dark. Experimental evidence for band-edge movement is based on comparing the dependence of V_{OC} on the photo-induced electron density in the absence and in the presence of an adsorbent. Because the photoelectron density is held constant, a change in V_{OC} can only happen if the bands move. The black line shows the results for a DSSC in the absence of an adsorbent. It can be seen that cografting chenodeoxycholate with the sensitizing dye N719 onto TiO_2 shifts the conduction-band-edge upward by 80 mV, which represents an 80-mV gain in the photovoltage. In contrast, the adsorbent guanidinium causes the conduction-band edge to move downward by 100 mV, which represents a 100-mV loss in photovoltage.

Figure 11 displays the effect of chenodeoxycholate and guanidinium on J_{SC} over a range of photoelectron densities at open circuit. In the absence of recombination at short circuit, to a very good approximation, and at the same light intensity, $J_{SC} = J_r$ at open circuit [36, 37].

$$J_{SC} = J_{inj} = |J_r| = -kc_{ox}n^\gamma, \qquad (4)$$

where J_{inj} is the charge injection density, c_{ox} is the concentration of the oxidized redox component, k the rate constant, which is related to the microscopic mechanism of recombination, and the exponent γ has typical values of 2–3 [16, 48] but can be larger [49]. The values of J_{inj} at short circuit and open circuit are assumed to be the same. Experimental evidence for recombination is based on comparing the dependence of the recombination current density on the electron density in the absence and presence of an adsorbent. Because the photoelectron density is held constant, a shift of the curve indicates a change in the rate constant k for recombination. Exposing the TiO_2 surface to chenodeoxycholate is seen to increase the recombination current density by fivefold. In contrast, guanidinium decreases the recombination current density by 20-fold.

Figure 12 shows that in the case of chenodeoxycholate, the collective effect of both the 80-mV upward band-edge shift and fivefold faster recombination resulted in a 25 mV increase in photovoltage [37]. This implies that increasing recombination

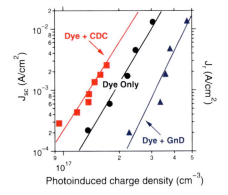

Fig. 11 Effect of chenodeoxycholate and guanidinium on J_{SC} as a function of the photoinduced charge density at open circuit. J_{SC} determines the recombination current density (J_r) at open circuit (4). Adapted with permission from [37] and [39]. Copyright 2005 and 2006, American Chemical Society

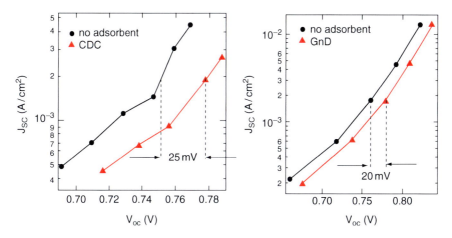

Fig. 12 Effect of chenodeoxycholate (*left*) and guanidinium (*right*) on the open-circuit photovoltage over a range of short-circuit photocurrent densities. Reprinted with permission from [37] and [39]. Copyright 2005 and 2006, American Chemical Society

led to a 55-mV loss of the possible photovoltage that would have been gained from just band-edge movement. For guanidinium, the collective effect of both the 100-mV downward band-edge shift and 20-fold slower recombination produced a similar improvement in the photovoltage of about 20 mV. This suggests that surface passivation resulted in a 120-mV voltage gain, which more than offset the loss of photovoltage from the downward band-edge movement. Thus, although both molecules improved the cell photovoltage, they do it by different mechanisms.

5 Conclusions

These studies show that creating more order in nanostructured architectures and judiciously modifying the surface properties of nanoporous films can significantly influence mechanistic factors that govern transport, recombination, and band energetics, which, in turn, determine the conversion efficiency of sensitized solar cells. Gaining a detailed understanding of these factors is critical to making rapid advances and improving cell performance.

Acknowledgments The author is grateful to Dr. Kai Zhu (National Renewable Energy Laboratory) for his help in organizing the figures and references. This work was supported by the Division of Chemical Sciences, Geosciences, and Biosciences, Office of Basic Energy Sciences and the Division of Photovoltaics, Office of Utility Technologies, US Department of Energy, under contract No. DE-AC36-08GO28308.

References

1. Nazeeruddin MK, Kay A, Rodicio I et al (1993) J Am Chem Soc 115:6382
2. Kay A, Grätzel M (1993) J Phys Chem 97:6272
3. Nazeeruddin MK, Pechy P, Renouard T et al (2001) J Am Chem Soc 123:1613
4. He JJ, Benko G, Korodi F et al (2002) J Am Chem Soc 124:4922
5. Hasobe T, Imahori H, Fukuzumi S et al (2003) J Mater Chem 13:2515
6. Plass R, Pelet S, Krueger J et al (2002) J Phys Chem B 106:7578
7. Blackburn JL, Selmarten DC, Nozik AJ (2003) J Phys Chem B 107:14154
8. Yu PR, Zhu K, Norman AG et al (2006) J Phys Chem B 110:25451
9. Chiba Y, Islam A, Watanabe Y et al (2006) Jpn J Appl Phys Lett (Express Letter) 45:L638
10. Basic Research Needs for Solar Energy Utilization: Report on the basic energy sciences workshop on solar energy ulitization, 18–21 Apr 2005. http://www.sc.doe.gov/bes/reports/files/SEU_rpt.pdf
11. Park NG, van de Lagemaat J, Frank AJ (2000) J Phys Chem B 104:8989
12. Nakade S, Saito Y, Kubo W et al (2003) J Phys Chem B 107:8607
13. Benkstein KD, Kopidakis N, van de Lagemaat J et al (2003) J Phys Chem B 107:7759
14. van de Lagemaat J, Benkstein KD, Frank AJ (2001) J Phys Chem B 105:12433
15. Cass MJ, Qiu FL, Walker AB et al (2003) J Phys Chem B 107:113
16. Kopidakis N, Schiff EA, Park NG et al (2000) J Phys Chem B 104:3930
17. Dittrich T, Lebedev EA, Weidmann J (1998) Phys Status Solidi A Appl Res 165:R5
18. Forro L, Chauvet O, Emin D et al (1994) J Appl Phys 75:633
19. Cao F, Oskam G, Meyer GJ et al (1996) J Phys Chem 100:17021
20. de Jongh PE, Vanmaekelbergh D (1996) Phys Rev Lett 77:3427
21. Nelson J, Haque SA, Klug DR et al (2001) Phys Rev B 6320:205321
22. Schlichthörl G, Park NG, Frank AJ (1999) J Phys Chem B 103:782
23. Dloczik L, Ileperuma O, Lauermann I et al (1997) J Phys Chem B 101:10281
24. Solbrand A, Lindstrom H, Rensmo H et al (1997) J Phys Chem B 101:2514
25. Bisquert J, Vikhrenko VS (2004) J Phys Chem B 108:2313
26. Macak JM, Tsuchiya H, Ghicov A et al (2005) Electrochem Commun 7:1133
27. Mor GK, Shankar K, Paulose M et al (2006) Nano Lett 6:215
28. Wang H, Yip CT, Cheung KY et al (2006) Appl Phys Lett 89:023508
29. Zhu K, Neale NR, Miedaner A et al (2007) Nano Lett 7:69
30. Wang ZS, Kawauchi H, Kashima T et al (2004) Coord Chem Rev 248:1381
31. Ferber J, Luther J (1998) Sol Energy Mater Sol Cells 54:265
32. Palomares E, Clifford JN, Haque SA et al (2003) J Am Chem Soc 125:475
33. Chen SG, Chappel S, Diamant Y et al (2001) Chem Mater 13:4629
34. Wang P, Zakeeruddin SM, Comte P et al (2003) J Phys Chem B 107:14336
35. Wang P, Zakeeruddin SM, Humphry-Baker R et al (2003) Adv Mater 15:2101
36. Schlichthörl G, Huang SY, Sprague J et al (1997) J Phys Chem B 101:8141
37. Kopidakis N, Neale NR, Frank AJ (2006) J Phys Chem B 110:12485
38. Zhu K, Vinzant TB, Neale NR et al (2007) Nano Lett 7:3739
39. Neale NR, Kopidakis N, van de Lagemaat J et al (2005) J Phys Chem B 109:23183
40. Park NG, Schlichthorl G, van de Lagemaat J et al (1999) J Phys Chem B 103:3308
41. van de Lagemaat J, Frank AJ (2001) J Phys Chem B 105:11194
42. van de Lagemaat J, Park NG, Frank AJ (2000) J Phys Chem B 104:2044
43. Frank AJ, Kopidakis N, van de Lagemaat J (2004) Coord Chem Rev 248:1165
44. Ruan CM, Paulose M, Varghese OK et al (2005) J Phys Chem B 109:15754
45. Perez-Blanco JM, Barber GD (2008) Sol Energy Mater Sol Cells 92:997
46. Shankar K, Mor GK, Fitzgerald A et al (2007) J Phys Chem C 111:21
47. Zhu K, Kopidakis N, Neale NR et al (2006) J Phys Chem B 110:25174
48. Kambili A, Walker AB, Qiu FL et al (2002) Physica E 14:203
49. Kopidakis N, Benkstein KD, van de Lagemaat J et al (2003) J Phys Chem B 107:11307

Implications of Interfacial Electronics to Performance of Organic Photovoltaic Devices

M.F. Lo, T.W. Ng, M.K. Fung, S.L. Lai, M.Y. Chan, C.S. Lee, and S.T. Lee

Contents

1 Introduction 170
2 Importance of Interfacial Studies in OPV Devices 170
3 Conceptual Aspects of Organic–Organic Heterojunction 172
4 Experimental Techniques and Sample Preparation 174
5 Ambient Effects on Copper Phthalocyanine/Fullerene Photovoltaic Interface ... 176
    5.1 UPS Studies on ITO/CuPc/C_{60} (Without Exposure) 176
    5.2 UPS Studies on ITO/CuPc/C_{60} (with Exposure) 177
    5.3 Effect of Exposure to Low Vacuum 178
6 The Role of Deposition Sequence and the Substrate Effect on C_{60}/CuPc and CuPc/C_{60}
    Heterojunctions 180
    6.1 The Role of Deposition Sequence (CuPc/C_{60} vs. C_{60}/CuPc) 181
    6.2 The Role of Substrate Work Function 183
    6.3 Summary on CuPc–C_{60} Interface 189
7 Metal-Doped Organic Layer as an Exciton Blocker/Optical Spacer 189
    7.1 Structural Considerations of Combining Optical Spacer and EBL 189
    7.2 The Limitation on Combining Optical Spacer
         and Exciton Blocker 190
    7.3 Effect of Doping the EBL with Metal 191
    7.4 Interface Study on Combined Optical Spacer
         and Exciton Blocker 193
8 Conclusion 195
References 195

Abstract It is widely known that electronic structures of metal/organic semiconductors and organic–organic interfaces have significant influences on performance of organic light-emitting devices. However, relatively little works have been done

---

M.F. Lo, T.W. Ng, M.K. Fung, S.L. Lai, C.S. Lee, (✉) and S.T. Lee
Center of Super-Diamond and Advanced Films (COSDAF), Department of Physics and Materials Science, City University of Hong Kong, Hong Kong, China
e-mail: apcslee@cityu.edu.hk

M.Y. Chan
Department of Chemistry, The University of Hong Kong, Pokfulam Road, Hong Kong, China

on their influences on organic photovoltaic devices. In this chapter, effects of deposition condition, deposition sequence, and substrate work function in controlling the energy level alignment between copper phthalocyanine and fullerene are introduced. In addition, effects of metal doping on the exciton blocker/cathode interface are summarized. Implications of these changes in interfacial electronic structures to the performance of OPV devices are discussed.

1 Introduction

Since the discovery of photo-charge generation from organic materials, much research efforts have been motivated to develop organic photovoltaic (OPV) devices due to their merits including low-cost and easy fabrication over a large and flexible area, etc. In the past couple of decades, up to ~5% of energy conversion efficiency has been achieved with either a donor–acceptor (D–A) bilayer structure or a blended bulk heterojunction structure [1, 2].

Many approaches have been attempted to improve performance of OPV devices. These include the applications of multiheterojunctions [3, 4] and highly efficient photoactive materials [5]. As in all other organic electronic devices, electronic structures of interfaces between various active layers have important influences on the performance of OPV devices. For example, energy offset between the highest occupied molecular orbital (HOMO) of the donor and the lowest unoccupied molecular orbital (LUMO) of the acceptor (HOMO$_D$ − LUMO$_A$) is widely recognized to define the maximum open-circuit voltage (V_{OC}), which limits the power conversion efficiency (PCE) of typical OPV devices. It is thus of interest to study the factors that control the energy level alignment in each individual interface with the aim to enable designs of more stable and efficient devices. In this chapter, we will summarize our recent works on interfacial studies on critical interfaces in a standard OPV device. Implications of these interfaces on device performance are discussed.

2 Importance of Interfacial Studies in OPV Devices

Initial efforts on developing OPV device using copper phthalocyanine (CuPc) and perylene teracarboxylic derivative in the form of a donor–acceptor bilayer heterojunction was first reported by Tang in 1985 [6]. As shown in Fig. 1a, the device consists of a donor–acceptor organic bilayer sandwiched by a high work-function anode (e.g., indium-doped tin oxide) and a relatively low work-function cathode (e.g., Al).

The PV process is initiated by the absorption of a photon in the donor (or the acceptor), leading to the formation of an exciton as shown in Fig. 1b. This is an electron–hole pair in an excited state, which is generated by electronic transition from the π-HOMO to the π*-LUMO of the donor material upon photo-excitation. The quasi-particle exciton will then settle in the state (E_{ex}) slightly above the HOMO and below the LUMO of the donor materials. The corresponding exciton

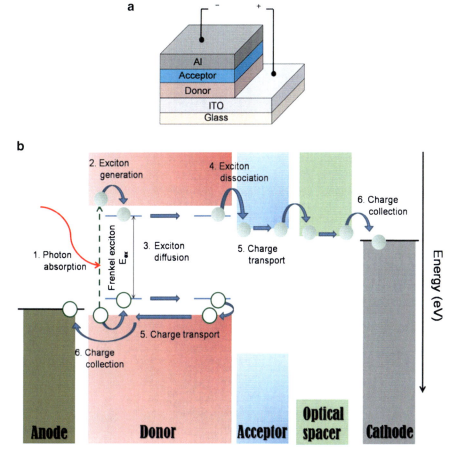

Fig. 1 A typical OPV device configuration and its working principle. (**a**) A standard bilayer device with donor–acceptor sandwiched between two electrodes; and (**b**) a schematic diagram showing operation mechanisms of an organic photovoltaic device

binding energy (E_B, ~0.2 eV in typical organic materials) can be estimated as the energy difference between the bandgap (E_g) of the donor and the E_{ex}. These excitons generated have a finite lifetime and would diffuse in the donor and acceptor layer. Those excitons diffuse to the donor–acceptor interface would experience an electric field, which would try to separate the electron–hole pairs. If the excitons do dissociate, the free electrons and holes would move, respectively, in the acceptor and donor toward the opposite electrodes leading to current output.

In recent studies, a thin optical spacer (also called an exciton blocker) is often inserted between the acceptor and the cathode. This layer, on one hand, modulates the interferential constructive maximum of incident light intensity close to the donor acceptor interface to maximize efficient photon absorption and charge dissociation. On the other hand, it also blocks excitons from reaching the cathode to suppress quenching and thus increases the chance of exciton dissociation.

3 Conceptual Aspects of Organic–Organic Heterojunction

Although typical organic materials used in OPV devices have energy gaps around 2 eV, V_{OC} of typical devices are about 0.5 V only [7]. In fact, the low V_{OC} is considered as a major factor that limits efficiency of OPV devices [8]. According to the classical thin-film solar cell concept or the metal–insulator–metal (MIM) model, the V_{OC} is simply equal to the difference between work functions of the two electrodes [9]. However, the validity of this model is debatable in devices based on fullerene (C_{60}) and its derivatives as they often show strong Fermi level pinning at the C_{60}/metal cathode interface [10]. By considering the Marcus theory for electron transfer, and the Schottky–Mott model (i.e., a common vacuum level for all materials) for estimating energy level alignments in organic semiconductors as shown in Fig. 2a, Rand et al. suggested that the maximum possible value of V_{OC} can be estimated as $|HOMO_D - LUMO_A| - E_B$ [7]. Scharber et al. have performed a systematic study on the relationship between V_{OC} and the energy levels of the donor–acceptor [10, 11]. Their results show that the V_{OC} follows (1):

$$V_{OC} = \frac{HOMO_D - LUMO_A}{e} - 0.3. \qquad (1)$$

It should be noted that the $HOMO_D - LUMO_A$ energy offsets are typically estimated with an implicit assumption of a common vacuum level for all layers in the devices. Unfortunately, numerous studies with photoelectron spectroscopies

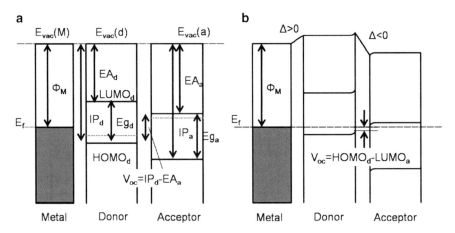

Fig. 2 Schematic representation of interfacial energy level diagram of a metal–donor–acceptor (**a**) with common vacuum level (E_{vac}) obeying the Schottky–Mott model, and (**b**) with discontinuities in the E_{vac} at the interfaces

and Kelvin probe measurements demonstrate deviation (Fig. 2b) from this common vacuum level picture. In the case of metal/organic junctions, interfacial dipole (Δ) up to 1 eV are not uncommon [12–16]. Due to the much weaker electronic interaction between organic molecules, organic–organic interfaces can, in principle, be described by the common vacuum level model. However, our previous work (see Fig. 3) has shown that a strong interfacial dipole (~0.4 eV) can also exist at some organic–organic heterojunctions, e.g., at the CuPc/F_{16}CuPc interface [17].

This clearly reveals that the common vacuum level assumption might not be valid in organic heterojunctions. In fact, at junctions involving donor/acceptor pairs, electronic interaction between the donor and the acceptor is expected to be large and naturally leads to deviation from the common vacuum scenario. This suggests that the $HOMO_D - LUMO_A$ energy offset at the donor–acceptor interface cannot be simply estimated from the difference between the ionization potential (IP) of the donor and electron affinity (EA) of the acceptor. Energy level bending and interfacial dipole should thus be considered for estimating the $HOMO_D - LUMO_A$ energy offset.

Despite the strong Fermi level pinning at the C_{60}/cathode interfaces, the work function of anode is also considered to influence the overall V_{OC} of OPV devices. For example, Bhosle et al. found that devices fabricated on GaZnO substrate, which has a work function lower than ITO, could yield a higher V_{OC} [18]. On the other hand, Hong et al. showed that devices fabricated on ITO of lower work functions in general have V_{OC} higher than those fabricated on ITO of higher work functions [19]. On one hand, these observations are inconsistent with the $HOMO_D - LUMO_A$ energy offset model, which suggests that the V_{OC} is independent of the electrode work function. On the other hand, the results cannot be explained by the MIM model, which suggests an opposite trend.

Recently, our works show that electronic structures of some organic–organic interfaces such as tris(8-hydroxyquinoline)aluminum (Alq_3)/CuPc can be

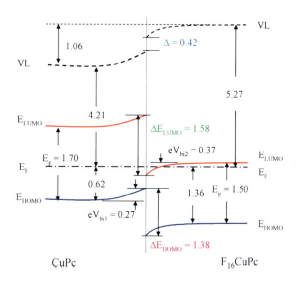

Fig. 3 A schematic interfacial energy level diagram showing the existence of a strong interfacial dipole (0.42 eV) at the CuPc/F_{16}CuPc organic–organic heterojunction. Reprinted with permission from [17]. Copyright 2006, American Institute of Physics

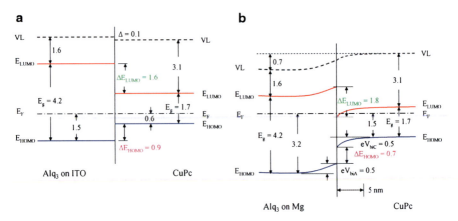

Fig. 4 Schematic energy level diagrams of the CuPc/Alq$_3$ junction on (**a**) ITO and (**b**) Mg substrates. All the values shown are in units of eV. Reprinted with permission from [20]. Copyright 2006, American Institute of Physics

substantially influenced by the substrate work function as shown in Fig. 4 [20–22]. These results suggest that the substrate can play a crucial role in determining the energy level alignment in organic heterojunctions. Implications of this effect on OPV devices are discussed in Sect. 6.

The following section (Sect. 4) begins with a brief description on the experimental aspects of the interfacial studies. Principle of the analytical methods used in current studies [X-ray photoemission spectroscopy (XPS) and ultraviolet photoemission spectroscopy (UPS)] can be found in the literature [13]. In Sect. 5, we present studies on the ambient effects and the degradation mechanisms of the CuPc/C$_{60}$ interface using photoemission spectroscopic techniques. In Sect. 6, the substrate effects on the interfacial properties of the CuPc/C$_{60}$ and C$_{60}$/CuPc heterojunctions are discussed. The electronic properties and the electron conduction pathway at the C$_{60}$/optical spacer interface are discussed in Sect. 7.

4 Experimental Techniques and Sample Preparation

All photoemission studies described here were performed in a VG ESCALAB 220i-XL ultrahigh vacuum (UHV) photoemission spectroscopy system with a base vacuum up to 10^{-10} Torr. A monochromatic aluminum Kα X-ray source was used for providing photons with 1486.6 eV energy for XPS analysis. Unfiltered He I (21.22 eV) and He II (40.8 eV) photons from a gas-discharge lamp were used as excitation sources for UPS studies. The instrumental energy resolution is 90 meV as estimated from the Fermi edge of a cleaned Au film. The photoelectrons generated by the photoemission process were collected with a concentric hemispherical analyzer (CHA) equipped with six channeltrons. Figure 5 shows a simplified schematic drawing of the surface analysis system.

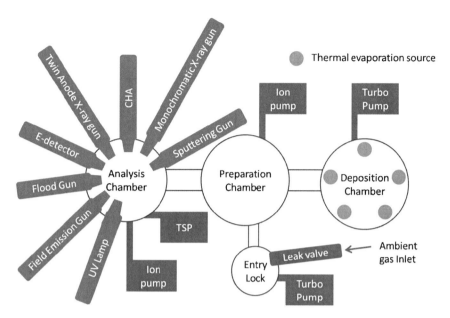

Fig. 5 A schematic drawing of the VG ESCALAB 220i-XL photoemission spectroscopy system

The system consists of three interconnected chambers; namely the analysis chamber, which is equipped with all the instruments for various surface analytical analysis; a multiport carousel chamber attached with a fast entry sample load-lock; and a deposition chamber. The sample entry load-lock is connected with an inlet for controllable ambient gas leakage which allows experiments involving controlled exposure to various atmospheres. The base pressures of the three chambers are 3×10^{-10} Torr, 5×10^{-10} Torr, and 2×10^{-10} Torr, respectively.

Unless mentioned otherwise, ITO-coated glass was used as substrates in all experiments. Before loading the ITO-coated glass substrates into the UHV system, they were thoroughly cleaned using Decon 90 and rinsed in deionized water, then dried in an oven for at least 2 h, and finally treated in an ultraviolet (UV) ozone chamber for 25 min. UPS spectra were recorded with a negative sample bias of 5.0 V with respect to ground in order to allow observation of electrons with the lowest kinetic energy. Materials used including CuPc (from Luminescence Technology Corp. and sublimated twice) and C_{60} (from American Dye Source, Inc.; purity > 99.95%), bathophenanthroline (BPhen), ytterbium (Yb), CsF, MoO_3, and Mg were evaporated in the deposition chamber using tantalum evaporation cells. Deposited film thickness was monitored with a quartz crystal microbalance. Freshly prepared samples were transferred, without vacuum break, to the analysis chamber for photoemission measurements.

OPV devices were also fabricated and characterized for comparisons with the surface analysis results. Patterned ITO-coated glass substrates with a sheet resistance of 30 Ω per square were used as anodes of the OPV devices. Organic materials and metal films were prepared by thermal evaporation at a vacuum of 10^{-6} Torr with evaporation rates controlled at 1–2 Å s^{-1}. Current density–voltage (J–V) characteristics were measured in the dark and under AM 1.5 solar illumination from an Oriel 150 W solar simulator using a programmable Keithley model 237 power source. All the measurements were carried out in air at room temperature.

5 Ambient Effects on Copper Phthalocyanine/Fullerene Photovoltaic Interface

While the energy conversion efficiency of OPV devices is continuously improved, device instability remains a major hurdle for practical usage. Ambient environment containing moisture and oxygen is believed to be deleterious to most organic semiconductors, leading to aging and degradation of photovoltaic cell [23–25]. Seki et al. have recently reported that oxygen behaves as a p-type dopant in C_{60} films [26]. Nevertheless, the impact of ambient gas on the V_{OC} is yet unclear.

As discussed in the previous section, performance of OPV devices depends greatly on the electronic structures at the donor–acceptor interfaces around which photoactivity takes place [27, 28]. Specifically, the theoretical limit of V_{OC} in OPV device has been shown to be linearly dependent on the $HOMO_D - LUMO_A$ energy offset [11]. Here, our focus is on how exposure to ambient air would influence the interface. We demonstrate that the $HOMO_{CuPc}-LUMO_{C60}$ energy offset (theoretical maximum V_{OC}) of OPV increases from 0.64 to 0.81 eV upon exposure to a lower vacuum. The findings offer a reasonable explanation to the initial increase in V_{OC} during initial aging of OPV devices [27]. Further, the results also show that exposure to ambient gases can reduce carrier density and conductivity of C_{60}.

5.1 UPS Studies on ITO/CuPc/C_{60} (Without Exposure)

Figure 6 shows evolution of UPS spectra of increasing C_{60} film thickness on CuPc. The bottom spectrum was measured from a 100 Å thick pristine CuPc layer deposited on an ITO substrate. After each UPS measurement, C_{60} film was incrementally deposited onto CuPc. Throughout the experiment, the sample was kept in vacuum of $< 2 \times 10^{-9}$ Torr (the lowest vacuum during C_{60} deposition), and the whole experiment was completed within 4 h to minimize contamination.

As C_{60} coverage increases, the high-binding energy cutoff position shows little changes (Fig. 6a). At the same time, the HOMO features of CuPc are gradually replaced by those of C_{60} (Fig. 6b). Again, the C_{60} HOMO position shows only a small change (about 0.05 eV) upon increasing thickness of C_{60} to 100 Å. These negligible shifts in both the secondary cutoff region and the onset of HOMO peak suggest that the C_{60}/CuPc interface can be well described by the classical Schottky–Mott model with a common vacuum level and flat energy levels. The IPs

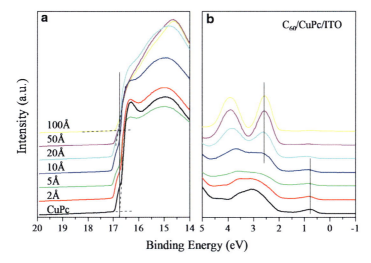

Fig. 6 UPS spectra showing (**a**) the secondary electron cutoff, and (**b**) the HOMO region near the Fermi level as a function of increasing C_{60} deposition in UHV. Zero binding energy corresponds to the Fermi level of the substrate. Reprinted with permission from [31]. Copyright 2009, American Institute of Physics

obtained for CuPc and C_{60} are 4.85 eV and 6.41 eV, respectively, in good agreement with the previous results [21].

5.2 UPS Studies on ITO/CuPc/C_{60} (with Exposure)

The above experiment was then repeated by adding an "exposure" step after each C_{60} deposition. After each UPS measurement of the pristine C_{60} film on a CuPc film (100 Å) on ITO, the sample was transferred to the preparation chamber for exposure to a lower vacuum of 10^{-5} Torr for 1 h and then reloaded into the analysis chamber for another UPS measurement. Figure 7 shows the measured UPS spectra. Spectra labeled "x Å" and "x Å-ex" denote the UPS spectra of the samples with a total C_{60} thickness of x Å measured before and after the "exposure" step.

Upon deposition of C_{60}, characteristic features of CuPc are progressively attenuated, accompanied by an obvious shift in the high binding energy cutoff (Fig. 7a). During the first few deposition steps (< 4 Å), the E_{vac} shifted toward the high binding energy side by about 0.2 eV; then moved toward the low binding energy by 0.30 eV upon further C_{60} deposition. Detailed interfacial structure can be revealed by a careful study of the UPS spectra near the Fermi level as shown in Fig. 7b. The initial 0.2 eV downward shift in the E_{vac} as shown in Fig. 7a is accompanied with a shift in the HOMO position of CuPc. The UPS spectra measured with C_{60} coverage below 4 Å actually reflected the changes in the features of CuPc. As the C_{60} coverage further increases, the values of the E_{vac} and the HOMO peak position of

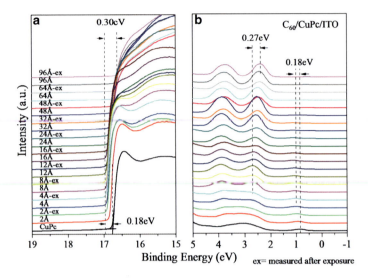

Fig. 7 Evolution of UPS spectra showing (**a**) the secondary electron cutoff, and (**b**) the HOMO region near the Fermi level as a function of increasing coverage of C_{60} deposited on CuPc (100 Å)/ITO before and after exposure to 10^{-5} Torr for an hour. Reprinted with permission from [31]. Copyright 2009, American Institute of Physics

C_{60} show an overall increase of 0.30 and 0.27 eV, respectively. The IPs of CuPc and C_{60} obtained are 4.90 eV and 6.42 eV, respectively.

5.3 Effect of Exposure to Low Vacuum

Changes in the E_{vac} and the HOMO peak positions of the C_{60} for the two experiments are summarized in Fig. 8. For the C_{60}/CuPc junction prepared in UHV condition, only small changes (about 0.10 and 0.05 eV, respectively) in the vacuum and the HOMO levels were observed at the interface region, as shown in Fig. 8a. In contrast, the C_{60}/CuPc junction exposed to 10^{-5} Torr shows very different electronic structures. The E_{vac} first decreases by about 0.25 eV at the interface region and then gradually increases as the thickness of C_{60} increases (Fig. 8b). On the other hand, the HOMO level shows an increasing trend. In Fig. 8b, the changes of energy level after each C_{60} deposition and exposure steps are shown by arrows with dashed and solid lines, respectively.

Results of the UPS analysis are summarized as schematic energy level diagrams in Fig. 9. LUMO levels are determined by subtracting the charge transport gaps of 1.9–2.3 eV for CuPc and C_{60}, respectively, from their HOMO levels [29].

Clearly, interfacial electronic structures of the C_{60}/CuPc junction prepared in UHV and 10^{-5} Torr of ambient-air show obvious differences. While the energy level diagram of the C_{60}/CuPc junction prepared under UHV (Fig. 9a) can be well

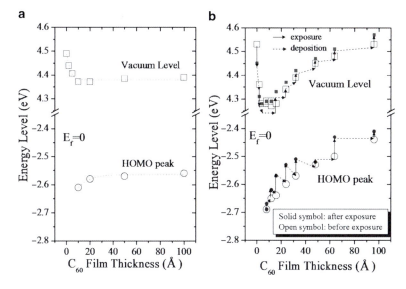

Fig. 8 Shifts of vacuum level and HOMO peak position of C_{60} film layer at C_{60}/CuPc heterojunction (**a**) without any exposure; and (**b**) before and after exposure to the controlled ambient air. *Open* and *solid symbols* correspond to measurements after the exposure and deposition steps. Reprinted with permission from [31]. Copyright 2009, American Institute of Physics

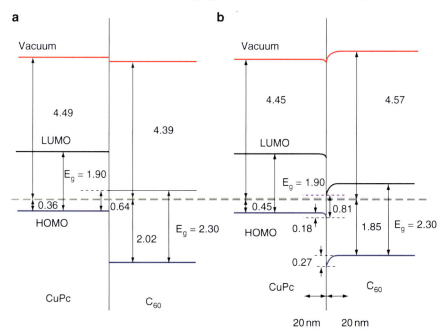

Fig. 9 Schematic energy level diagrams of C_{60}/CuPc heterojunction; (**a**) prepared under UHV condition, and (**b**) under exposure to a low vacuum of 10^{-5} Torr. All values are in eV. Reprinted with permission from [31]. Copyright 2009, American Institute of Physics

described by conventional nearly "common vacuum level" (dipole at the interface is only about 0.1 eV) and "flat energy levels" picture, the junction prepared at low vacuum shows obvious energy level bending.

UPS spectra of the C_{60} film show that the high energy electron cutoff shifts toward lower binding energy. It indicates that the "exposed" C_{60} film is p-typed doped by the ambient gas molecules. Moreover, the HOMO peaks of the C_{60} films are slightly broadened such that the highest occupied state (onset of HOMO peak) is slightly extended toward the Fermi level (Fig. 7b).

The space charges generated due to exposure to low vacuum oxygen in the accumulation layer in C_{60} cannot be balanced upon the contact formation with CuPc. As a result, the energy level of CuPc shows a 0.18 eV downward bending, and a charge trapping region is formed at the interface of C_{60}/CuPc. It is interesting to note that the $HOMO_{CuPc} - LUMO_{C60}$ energy offset is increased from 0.64 to 0.81 eV.

A recent study revealed that aging of OPV devices in ambient conditions led to an initial increase of V_{OC} before degradation [24]. The increase of $LUMO_{C60} - HOMO_{CuPc}$ observed here is consistent with the increase in V_{OC}. On the other hand, upon exposure to the ambient gases, the LUMO of C_{60} is shifted away from the Fermi level from 0.28 to 0.45 eV. It indicates a substantial reduction in carrier density and thus the overall conductivity. This is consistent with the reduction in conductivity observed in C_{60}-based organic field-effect transistor upon exposure to ambient air [30].

It should be pointed out that as OPV devices are typically fabricated under 10^{-5}–10^{-6} Torr over a couple of hours, it can be reasonably expected that in a typical OPV device, the actual interfacial electronic structures should be something in between Fig. 9a and 9b.

6 The Role of Deposition Sequence and the Substrate Effect on C_{60}/CuPc and CuPc/C_{60} Heterojunctions

In the previous section, it is shown that the energy offset at $HOMO_{CuPc} - LUMO_{C60}$ in the CuPc/C_{60} heterojunction is sensitive to the deposition condition. In this section, the deposition sequence and the effect of substrate work-function in controlling the interfacial energy level alignment of CuPc–C_{60}-based OPV devices are discussed.

It would be shown that the energy offset at $HOMO_D - LUMO_A$ can be changed from 0.64 (C_{60} on CuPc) to 0.86 eV (CuPc on C_{60}) by reversing the deposition sequence. Furthermore, by changing the substrate work function from 2.81 to 5.07 eV, the $HOMO_D - LUMO_A$ offset can be further tuned from 0.86 to 1.27 eV. The results suggest that electrodes in OPV devices can have significant influences on the electronic structures and energy levels of the donor–acceptor interface, thus providing a viable means for performance enhancement.

6.1 The Role of Deposition Sequence (CuPc/C$_{60}$ vs. C$_{60}$/CuPc)

The evolution of He I UPS spectra of C$_{60}$ with increasing CuPc coverage on it is shown in Fig. 10. The bottom spectrum refers to a 10 nm thick pristine C$_{60}$ layer deposited on a UV–ozone treated ITO substrate (work function of 4.90 eV). Upon CuPc deposition, the characteristic spectral features of C$_{60}$ show progressive attenuation and are gradually replaced by those of CuPc. The IP obtained for C$_{60}$ and CuPc are 6.44 eV and 4.89 eV, respectively, which are in good agreement with the previous results [21].

With increasing CuPc coverage, the high binding energy cutoff of the UPS spectrum shows an obvious shift of 0.78 eV toward the higher binding energy; meanwhile, the HOMO peaks of C$_{60}$ and CuPc also show shifts in the same direction of 0.25 eV and 0.44 eV, respectively. This result suggests that the interface formation is associated with charge transfer forming a C$_{60}^{\delta-}$–CuPc$^{\delta+}$ dipole. XPS results (not shown) show that the core level peaks of the two materials have the same shift as the valence band features. This indicates that the shifts measured in UPS are associated with the energy level bending and not due to the effect of sample charging.

Figure 11a and 11b show the energy levels diagrams of C$_{60}$ on CuPc and CuPc on C$_{60}$, respectively [31]. The remarkable differences between the electronic

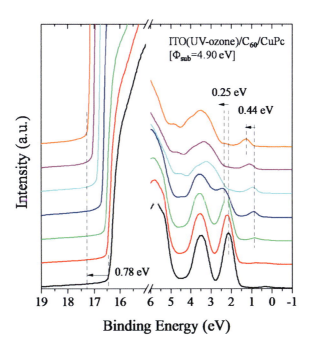

Fig. 10 Evolution of UPS spectra showing the secondary electron cutoff and the HOMO region as a function of CuPc deposited on a 10 nm C$_{60}$ on UV–ozone treated ITO substrate. Reprinted with permission from [35]. Copyright 2009, American Institute of Physics

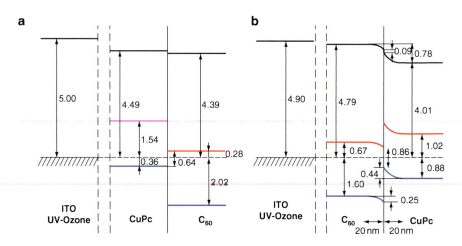

Fig. 11 A schematic energy level diagram of (**a**) C_{60} on CuPc heterojunction and (**b**) CuPc on C_{60} heterojunction deposited on UV–ozone-treated ITO substrates. Reprinted with permission from [35]. Copyright 2009, American Institute of Physics

structures of the two interfaces are attributed to the different extents of Fermi level pinning at the interfaces.

When CuPc is deposited on ITO (Fig. 11a), the vacuum level is spontaneously decreased to 4.49 eV due to the pinning of the HOMO of CuPc to the Fermi level. Under this circumstance, the charge density in such molecules would rearrange such that it is chemically and energetically favorable to transfer electrons to the underneath ITO layer upon contact formation. Because of this strong charge transfer at the ITO/CuPc junction, there is an abrupt E_{vac} decrease from 5.00 to 4.49 eV. Upon deposition of C_{60} onto CuPc, the vacuum aligns across the CuPc/C_{60} junction. Both the $HOMO_{CuPc}$ and the $LUMO_{C60}$ tend to pin to the metal Fermi level.

When C_{60} is deposited directly on ITO (Fig. 11b), C_{60} aligns itself by sharing a common E_{vac} with that of the ITO. Such vacuum level alignment is common for most wide bandgap organic materials where the Fermi level positions are far from both the HOMO and the LUMO levels [21]. However, upon deposition of the CuPc with an electron-donating property, the vacuum level is abruptly lowered by 0.80 eV due to transfer of charge carriers at the interface, leading to the formation of an interfacial dipole of 0.09 eV (Fig. 11b) and a significant band bending. These results suggest that the electronic structures at organic–organic interfaces are determined not only by the electronic coupling between the organic materials but also by the relative position of the Fermi level and the chemical potential equilibrium among the multilayer films.

At the same time, the energy offset at the $HOMO_D$–$LUMO_A$ has been effectively modified from 0.64 (C_{60} on CuPc) to 0.86 eV (CuPc on C_{60}) by reversing the deposition sequence. This implies that OPV devices with inverted structure should have higher V_{OC} than the conventional OPV devices. The improvement is mainly ascribed by the formation of the interfacial dipole at the C_{60}/CuPc heterojunction.

6.2 The Role of Substrate Work Function

Knowing the inverted configuration can provide a larger $HOMO_D - LUMO_A$ offset; here, we continue to discuss the role of substrate work function by using the inverted structure (C_{60}/CuPc) as an example. Similar UPS and XPS experiments were repeated by using substrates with work functions ranging from 2.81 to 5.05 eV (Table 1). Since all the UV–ozone- or oxygen plasma-treated samples (work function from 4.90 to 5.05 eV) gave similar results to that shown in Fig. 10, the UPS spectra from these samples are thus not shown here.

In contrast, the interface electronic structures do show qualitative differences when ITO work function is decreased below 4.17 eV. Figure 12 illustrates the evolution of He I UPS spectra of C_{60} with increasing CuPc coverage using hydrogen plasma-treated ITO substrate, which has a work function of 4.17 eV.

As shown in Fig. 12, the E_{vac} shows a small shift of 0.15 eV upon gradual deposition of CuPc; while the HOMO positions of both C_{60} and CuPc remain unchanged. The results suggest that energy levels are basically flat across the interfaces.

For the ITO substrates covered with CsF, results for two representative CsF thicknesses (2.5 and 6 nm) are shown in Figs. 13 and 14, respectively. Similar to the case for hydrogen plasma-treated ITO substrate, E_{vac} decreases upon CuPc deposition. The amount of E_{vac} shift increases with increasing CsF thickness (Figs. 13 and 14). On all substrates with CsF buffer layer, the HOMO levels of the organics show negligible changes upon interface formation.

Results of electronic properties extracted from Figs. 10, 12–14 are summarized as energy-level diagrams as shown in Fig. 15. As the main concern is on the C_{60}/CuPc interface, no UPS spectrum was measured as the C_{60} thickness increases. Therefore, the detailed interface formation processes at the substrate/C_{60} interfaces

Table 1 Work functions of various substrates

Treatment on ITO	Measured work function (eV)
Buffer layer	
ITO/CsF (6 nm)	2.81
ITO/CsF (2.5 nm)	3.19
ITO/CsF (1.5 nm)	3.68
ITO/CsF (0.6 nm)	4.19
Plasma treatment	
H_2 plasma treated	4.17
UV–ozone treated	4.90
O_2-plasma treated	5.05

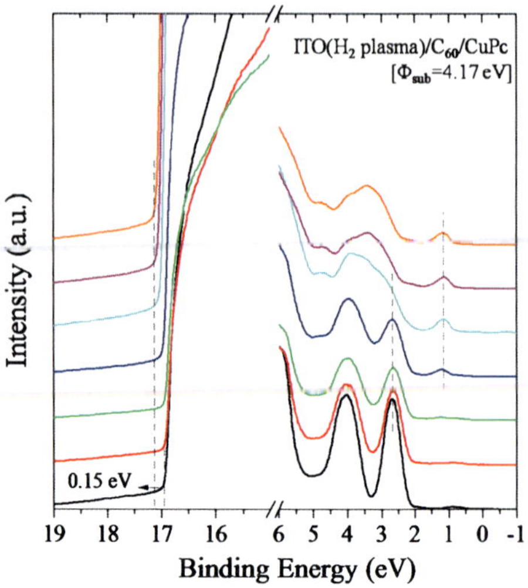

Fig. 12 Evolution of UPS spectra showing the secondary electron cutoff and the HOMO region as a function of CuPc deposited on a 10 nm C_{60} on H_2-plasma-treated ITO substrate, which has a work function of 4.17 eV. Reprinted with permission from [35]. Copyright 2009, American Institute of Physics

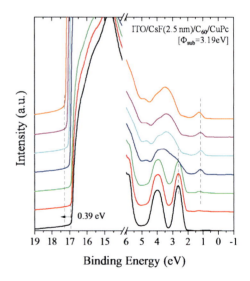

Fig. 13 Evolution of UPS spectra showing the secondary electron cutoff and the HOMO region as a function of CuPc deposited on a 10 nm C_{60} film on an ITO substrate covered with 2.5 nm of CsF. Reprinted with permission from [35]. Copyright 2009, American Institute of Physics

were not measured, and the energy levels at the substrate/C_{60} interfaces (as well as the ITO/CsF interfaces) are not shown.

For the UV–ozone-treated substrate with a work function of 4.90 eV, the C_{60} layer, as described by the Schottky–Mott model, follows the E_{vac} alignment with the substrate (Fig. 15a). Upon CuPc deposition, energy level bending and interface dipole can be clearly observed at the C_{60}/CuPc interface, suggesting that there is considerable charge transfer from CuPc to C_{60}. This can be understood by considering the interface formation process. Just before the two organics form contact,

Fig. 14 Evolution of UPS spectra showing the secondary electron cutoff and the HOMO region as a function of CuPc deposited on a 10 nm C_{60} film on an ITO substrate covered with 6 nm of CsF. Reprinted with permission from [35]. Copyright 2009, American Institute of Physics

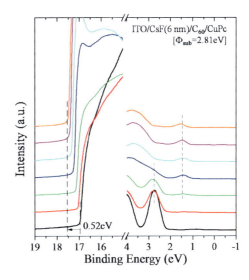

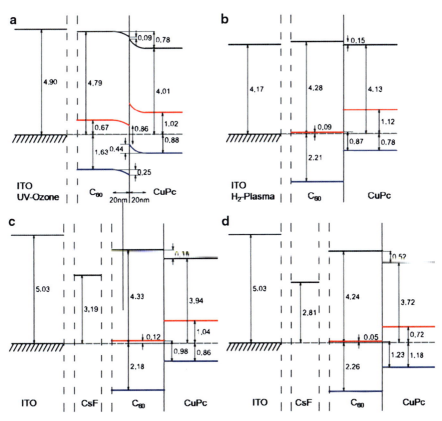

Fig. 15 Schematic energy level diagrams of CuPc on C_{60} heterojunction deposited on substrates of (**a**) UV–ozone-treated, (**b**) H_2-plasma-treated ITO, (**c**) ITO/CsF (2.5 nm), and (**d**) ITO/CsF (6 nm). All the values are in unit of eV. Energy levels between two *vertical dashed lines* were not measured. Reprinted with permission from [35]. Copyright 2009, American Institute of Physics

they share the same E_{vac}. As CuPc has an IP of ~4.9 eV, its HOMO level is thus close to the Fermi level (0.09 eV below the Fermi level), resulting in a high mobile carrier density in CuPc upon interface formation. This leads to spontaneous charge redistribution at the interface to achieve equilibrium and Fermi level matching. Consequently, there would be an electron accumulation layer in C_{60} and a depletion zone in adjacent CuPc. As a result, a heterojunction with energy level bending and dipole is formed as shown in Fig. 15a. A direct consequence of the dipole is that the $HOMO_D - LUMO_A$ offset, and thus the maximum possible V_{OC}, is increased.

When the substrate work function decreases to 4.2 eV, which is close to the EA of C_{60} (Fig. 15b) for a hydrogen plasma-treated ITO substrate, the LUMO of C_{60} would align with the Fermi level, leading to a common E_{vac} alignment at the ITO/C_{60} interface. Interface formation between CuPc and C_{60} also would result in a common E_{vac} as the IP of CuPc (~4.9 eV) is considerably larger than the work function of the underneath C_{60} (4.28 eV). Consequently, upon contact formation of CuPc with C_{60}, the Fermi level would be far from both the LUMO and the HOMO of CuPc. Hence, the amount of mobile charge carriers generated in CuPc is small, leading to flat energy level alignment as shown in Fig. 15b.

When the work function of the substrate further decreases from 4.17 to 2.81 eV by using CsF buffer layer of different thickness, the interfaces of C_{60}/CuPc show three general features (Figs. 15c and 15d) (1) substantial interface dipoles were observed at all interfaces and the E_{vac} alignment assumption is no longer valid, (2) the LUMO of C_{60} was aligned to the Fermi level, and (3) flat energy levels were observed at the C_{60}/CuPc interfaces.

In the interface formation process, the interface dipole and the Fermi level pinning at the CsF/C_{60} interface can be elucidated as follows. Before the contact formation, the E_{vac} of CsF and C_{60} are at the same level, and the LUMO of C_{60} is substantially below the Fermi level. Hence, when CsF and C_{60} come into contact, electron would spontaneously flow from the substrate (ITO/CsF) to the LUMO of C_{60}. This would induce an abrupt energy level offset at CsF/C_{60} interface until the LUMO of C_{60} is aligned to the Fermi level. Therefore, the E_{vac} alignment model is no longer valid. In Figs. 15c and 15d, the position of $LUMO_{C60}$ is located just at 0.12 and 0.05 eV above the Fermi level. The strong Fermi level pinning would generate a large number of mobile carriers in the C_{60} layer. When CuPc is further deposited onto the C_{60} layer, flat energy levels appear on both sides of the interface, implying there is little charge transfer across the C_{60}/CuPc interface.

On the other hand, the E_{vac} offsets suggest that a substantial interface dipole and charge transfer exist at the C_{60}/CuPc interface. This is in apparent contradiction to the observation of flat energy levels across the interface. It is considered that the strong Fermi level pinning effect of the LUMO of C_{60} might have caused a major rearrangement of electron cloud at the C_{60} surface so that the C_{60} surface behaved somewhat like a metal. In particular, the wave functions of mobile electrons in the LUMO of C_{60} would extend into the vacuum like a metal surface. When CuPc molecules are deposited onto the C_{60} surface, the "spilling" of the electron cloud out of the sample surface is pushed back into the C_{60} layer by the CuPc molecules, leading to a lowering of the vacuum level. This phenomenon, often referred to as the "pillow effect" in metal/organic contact, would lead to an abrupt E_{vac} change with almost flat energy levels [32, 33]. In addition, the dipole at the C_{60}/CuPc interface increases as the substrate work function decreases (i.e., with increasing CsF thickness). As the $HOMO_D - LUMO_A$ energy offset is directly coupled to the downward shift of the E_{vac} at the C_{60}/CuPc interface, it suggests that the theoretical maximum V_{OC} in the OPV device would increase with increasing CsF thickness.

The above results show that the energy level alignment at the C_{60}/CuPc interface can be considerably tuned by using substrates with different work functions. Changes in the LUMO level of C_{60} and the HOMO level of CuPc with the substrate work function (ϕ_{sub}) are shown in Fig. 16a. The variation of the work function of the C_{60} surface (ϕ_{C60}) is also shown in the inset of Fig. 16a. When ϕ_{sub} is above 4.2 eV, ϕ_{C60} varies linearly with ϕ_{sub} with a slope close to unity. When ϕ_{sub} is smaller than 4.2 eV, ϕ_{C60} remains constant at about 4.3 eV. It can also be seen from Fig. 16a that the LUMO level of C_{60} is independent of the substrate work function when it is smaller than 4.2 eV. It reflects the Fermi-level pinning of the energy levels of C_{60} when it is deposited on substrates with a low work function.

Figure 16a also shows that the variation of the HOMO level of CuPc is different from that of the LUMO level of C_{60}. The HOMO level of CuPc is almost constant when ϕ_{sub} is lower than 2.8 eV; it then increases as ϕ_{sub} increases. The vertical distance between the two lines in Fig. 16a is the $HOMO_D - LUMO_A$ offset, which is plotted in Fig. 16b. It shows that the $HOMO_D - LUMO_A$ offset can be considerably tuned between 0.86 and 1.27 eV by changing the substrate work function. In particular, the $HOMO_D - LUMO_A$ offset in the C_{60}/CuPc-based OPV devices follows roughly a U-shaped curve. The minimum point of the curve showing the offset at $HOMO_{CuPc} - LUMO_{C60}$ is located at a substrate work function of 4.2 eV, at which $LUMO_{C60}$ is pinned to the Fermi level of the substrate. This value is considered to be a specific threshold for the nonmetallic substrate to cause a Fermi level pinning of C_{60}. When the substrate work function is above this threshold value, the energy alignments of both C_{60} and CuPc are affected by the value of ϕ_{sub} according to their corresponding charge densities, and the $HOMO_{CuPc} - LUMO_{C60}$ energy offset increases with the value of ϕ_{sub}. However, when the value of ϕ_{sub} is below this specific threshold, the $LUMO_{C60}$ is pinned by the substrate's Fermi level; while the position of $HOMO_{CuPc}$ is continuously modified by the effect of the substrate. This is mainly attributed to the increased surface dipole induced at the C_{60}/CuPc interface. As a result, the $HOMO_{CuPc} - LUMO_{C60}$ energy offset across the interface increases when substrate has a ϕ_{sub} smaller than 4.2 eV.

These results clearly demonstrate that the energy offset of $HOMO_{CuPc} - LUMO_{C60}$ can be readily tuned from 0.86 to 1.27 eV by simply changing the work function of the ITO substrate via plasma treatments or inserting a thin buffer layer. As the $HOMO_{CuPc} - LUMO_{C60}$ offset is closely related to the maximum possible V_{OC} in an OPV device, the present results suggest that the V_{OC} of OPV device should potentially be tunable by varying the substrate work function.

While the present works suggest that V_{OC} and thus the performance of OPV devices would be tunable via the substrate effects, there are several limitations that

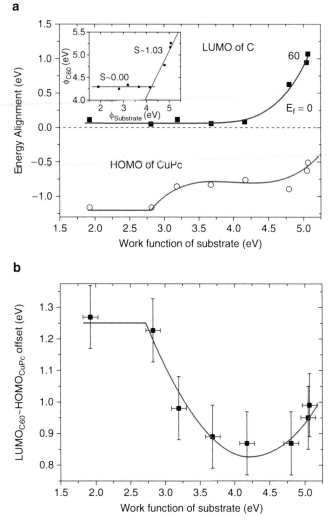

Fig. 16 A graph showing (**a**) the dependence of HOMO$_{CuPc}$ and LUMO$_{C60}$ with respect to substrate work function. The *inset* in (**a**) shows the work function of C$_{60}$ (Φ_{C60}) in relation to Φ_{sub}. (**b**) HOMO$_{CuPc}$–LUMO$_{C60}$ offset as a function of substrate work function. Reprinted with permission from [35]. Copyright 2009, American Institute of Physics

one should be aware of. First, it should be pointed out that the current results are obtained from nonmetallic substrates. On the other hand, it should be noted that while the HOMO$_D$ − LUMO$_A$ offset affects the maximum possible V_{OC}, experimental V_{OC} values obtained from typical OPV characterization are limited by numerous other factors. For example, to obtain the maximum possible V_{OC}, *I–V* measurements of an OPV device should be carried out at a temperature of ~175 K under an incident optical intensity of 150 mW cm^{-2} [6]. Furthermore, precaution should be also made to minimize the leakage current [34]. As these factors would reduce the V_{OC} to different degrees, care should be taken when comparing V_{OC} data in the literatures to the present results. Finally, as mentioned earlier, C_{60} is very sensitive to the vacuum condition during film deposition. Heterojunctions with C_{60} prepared in typical device fabrication conditions (e.g., vacuum of 10^{-5} Torr) can show considerable differences to those prepared at UHV ($<10^{-9}$ Torr). Nevertheless, the present work does demonstrate that the substrate effects could be potentially exploited to enhance the performance of OPV and other organic electronic devices via tuning interfacial energy level offsets [35].

6.3 Summary on CuPc–C_{60} Interface

Electronic structures of the interface between CuPc and C_{60} were found to be dependent on the deposition sequence. The $HOMO_D$ – $LUMO_A$ offset can be increased from 0.64 to 0.86 eV by using an inverted OPV structure (i.e., ITO/C_{60}/CuPc/metal) consistent with the observed increase in V_{OC} in the inverted OPV device. It appears that, in additional to the interactive coupling between the organics, the Fermi level position of the system also has important influences on the interfacial electronic structures. The $HOMO_D$ – $LUMO_A$ offset at the C_{60}/CuPc interface in the inverted structure (ITO/C_{60}/CuPc) can be changed from 0.86 to 1.27 eV via controlling the substrate work function by using CsF as a buffer layer or various surface treatments on the ITO substrate. These results suggest the possibility of using the substrate effects to enhance the performance of OPV devices.

7 Metal-Doped Organic Layer as an Exciton Blocker/Optical Spacer

For devices with a simple bilayer structure (ITO/CuPc/C_{60}/Al), it is believed that only a small fraction of the absorbed light contributes to the photocurrent while some of the generated excitons are quenched when they reach the metal electrode via diffusion. Therefore, an optical spacer or exciton blocking layer (EBL) has been introduced between the C_{60} and the metal electrode to improve OPV performance [3, 36–39]. Table 2 summarizes the definition, working principle, and basic requirements for the optical spacer and the exciton blocker.

7.1 Structural Considerations of Combining Optical Spacer and EBL

Despite the different functions and working principle of the optical spacer and the exciton blocker, they share the same physical location in an OPV device, that is, in

Table 2 A summary of the definition, working principle, and basic requirements for optical spacer and exciton blocker

	Definition	Principle	Basic requirements
Optical spacer	A highly transparent layer inserted between the acceptor layer and the cathode	To modulate the optical intensity in the device such that the intensity maximum would be close to the donor–acceptor interface where absorption and charge dissociation are efficient	High electrical conductivity High optical transmission, typically, 10–30 nm
Exciton blocker	A wide bandgap material inserted between the acceptor layer and the cathode	To block exciton diffusion into the cathode. It reduces exciton quenching at the cathode so that more excitons can contribute to the generation of electricity	A wide bandgap materials Low electrical resistance limited to 10 nm

between the acceptor layer and the cathode. This poses an interesting possibility that both the optical spacer and exciton blocker can be combined as one single layer provided that such a layer possesses the following properties:

1. A high EA [38, 40] and a high IP [3] to allow efficient electron injection and to effectively block the diffusion of exciton.
2. A low optical absorption coefficient across the solar spectrum and a high electrical conductivity such that it allows large flexibility in tuning the optimum thickness required for OPV cells.

Figure 17a shows a schematic representation of interferential wave interacting with the incident and reflecting light ray inside a device. The arrows represent modulate light intensity after reflection. Originally, the maximum of the interferential wave function is mismatched with the donor–acceptor interface where the exciton dissociates. This discrepancy cannot be fixed by changing the thickness of either the donor or acceptor layer due to the limited exciton diffusion length. However, with an optical spacer shown in Fig. 17b, the intensity maximum can be adjusted right at the donor–acceptor interface where absorption and charge dissociation are taking place.

7.2 The Limitation on Combining Optical Spacer and Exciton Blocker

Figure 18 shows the photovoltaic response (lines with empty circle symbols) of devices with a structure of ITO/CuPc/C_{60}/BPhen/Al as a function of BPhen layer

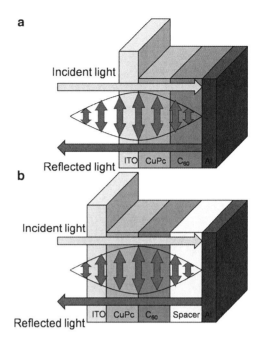

Fig. 17 Schematic representation of distribution of light within a bilayer device (**a**) without optical spacer, and (**b**) with an optical spacer

thickness under an illumination intensity of 45 mWcm^{-2}. For a simple structure without any optical spacer (ITO/CuPc/C$_{60}$/Al), the device exhibits a short-circuit current density (J_{SC}) = 4.0 mAcm^{-2}, V_{OC} = 0.31 V, fill factor (FF) = 0.22, and PCE = 0.63%. With an optimized BPhen thickness of 5 nm, the device exhibits a J_{SC} of 5.8 mAcm^{-2}, a V_{OC} of 0.34 V, a FF of 0.45, and a PCE of 1.99%.

However, the photovoltaic performance degrades sharply as the BPhen thickness increases. For example, FF and PCE rapidly drop to 0.22 and 0.003% for a BPhen thickness of 20 nm, which is attributed to the increased cell resistance as the EBL thickness exceeds the characteristic penetration depth of Al [37, 41].

7.3 Effect of Doping the EBL with Metal

Though metal-doped organic materials have been used as electron-transporting materials in organic light-emitting diodes [42, 43], their applications in OPV devices have not been fully explored. We demonstrate that photovoltaic performance is improved when Yb is doped into the BPhen layer. With a 5 nm thick Yb:BPhen layer, the device shows improved photovoltaic response; J_{SC} = 6.4 mA cm^{-2}, V_{OC} = 0.40 V, and FF = 0.45. This yields a high PCE of 2.57%. Interestingly, both FF and PCE remain at high values even when the Yb:BPhen thickness increases to 30 nm (lines with solid square symbols in Fig. 18). Clearly, Yb:BPhen functions not only as an exciton blocker but also as an optical spacer.

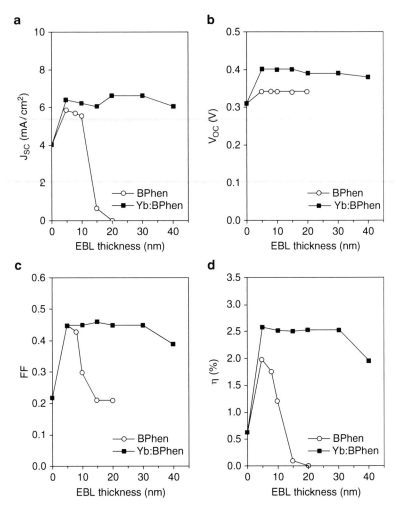

Fig. 18 Photovoltaic response of devices with structure of ITO/CuPc/C$_{60}$/EBL/Al as functions of EBL thickness, where EBL is either BPhen or Yb:BPhen. (Short-circuit current densities and PCE of all devices have been corrected for spectral mismatch factor of the detector.) Reprinted with permission from [45]. Copyright 2006, American Institute of Physics

spatially displacing the donor–acceptor interface to a better position to achieve maximum J_{SC} and PCE. The high FF is primarily a consequence of low series resistivity of the metal-doped thin film. Electrical conductivities of both the BPhen and the Yb:BPhen layers have been measured from organic layers sandwiched between an ITO anode and an Al cathode. Upon metal doping, the conductivity (σ) of BPhen film ($\sigma = 3.0 \times 10^{-5}$ S cm^{-1}) increases tenfold to 3.3×10^{-4} S cm^{-1}.

Figure 19 shows the absorption coefficients of BPhen and Yb:BPhen films on quartz substrates. For reference, the absorption coefficients of CuPc and C$_{60}$ are also

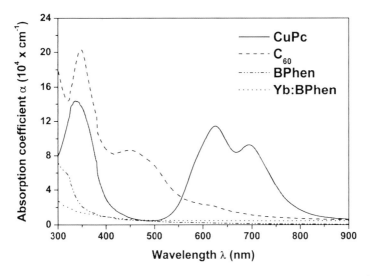

Fig. 19 Absorption coefficients of BPhen (*dashed–dotted line*) and Yb:BPhen (*dotted line*) on quartz substrates. For reference, absorption coefficients of CuPc (*solid line*) and C_{60} (*dashed line*) are also shown. Reprinted with permission from [45]. Copyright 2006, American Institute of Physics

shown. Clearly, the relatively low absorption coefficients within the solar spectrum for both the BPhen and Yb:BPhen films indicate their applicability as optical spacers. It is worthy to note that both BPhen and Yb:BPhen films show similar absorption edge at $\lambda = 354$ nm, giving the same optical bandgap E_g of 3.5 eV.

7.4 Interface Study on Combined Optical Spacer and Exciton Blocker

Figure 20a shows the evolution of the UPS spectra upon deposition of BPhen on C_{60} film on an ITO substrate. The bottom spectrum was measured from a 100 Å thick pristine C_{60} layer deposited on the ITO substrate. After each UPS measurement, BPhen film was incrementally deposited onto C_{60}. For the pristine C_{60} film, the HOMO edge is found at 1.56 eV with respect to the Fermi level, while the IP is measured to be 6.45 eV, which is in good agreement with the literature values [38]. Upon BPhen deposition, the entire UPS spectrum shifts gradually toward the higher binding energy. As BPhen coverage increases to 16 Å, the HOMO features of C_{60} are gradually replaced by those of BPhen. The peak of HOMO shifts gradually toward higher binding energy up to 0.37 eV, reflecting a molecular-level bending at the interface. Meanwhile, an interface dipole of 1.06 eV (by subtracting the total E_{vac} shift by that contribution from band bending) can be observed, implying

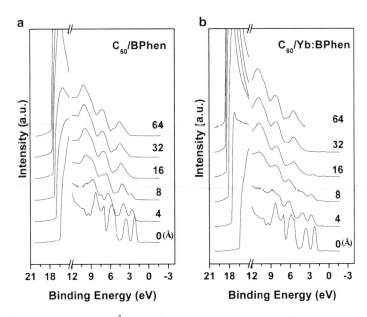

Fig. 20 UPS spectra of a 100 Å thick C_{60} film with (**a**) different BPhen overlayers, and (**b**) different Yb:BPhen overlayers. The coverage (Å) of either BPhen or Yb:BPhen layer is indicated. Reprinted with permission from [45]. Copyright 2006, American Institute of Physics

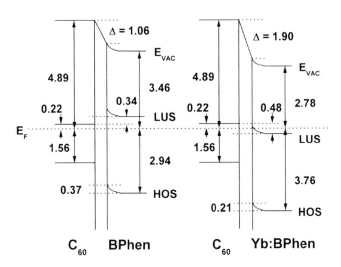

Fig. 21 Energy level diagrams of the C_{60}/BPhen and the C_{60}/Yb:BPhen contacts. Reprinted with permission from [45]. Copyright 2006, American Institute of Physics

a charge transfer process between the C_{60} and BPhen, which is due to the electron-accepting nature of C_{60}. The above experiment was repeated by using Yb:BPhen. The UPS spectra of C_{60} with varying coverage of Yb:BPhen overlayers are shown in Fig. 20b. Similarly, the entire UPS spectrum rigidly shifts toward higher binding energies with increasing Yb:BPhen thickness. The E_{vac} is lowered to a larger extent, giving an interfacial dipole of 1.90 eV, which may be attributed to both electron-accepting characteristic of C_{60} and the metal-doping effect [44].

Figure 21 shows schematic energy level diagrams for the C_{60}/BPhen and the C_{60}/Yb:BPhen interfaces. The positions of the LUMO were calculated from the optical bandgap of C_{60} (1.78 eV), BPhen, and Yb:BPhen as determined by optical absorption measurement in Fig. 19. From the UPS spectra, the IP of BPhen and Yb:BPhen are 6.40 and 6.54 eV, respectively. The LUMO – LUMO offset at the C_{60}/BPhen contact is +0.34 eV and is reduced to −0.48 eV at the C_{60}/Yb:BPhen contact. There is effectively no barrier for electrons to move from C_{60} to Yb:BPhen. This is responsible for the performance improvement of the OPV devices even with thicker Yb:BPhen layers [45].

8 Conclusion

In summary, possible influences of electrode on the V_{OC} of OPV devices are discussed. It is demonstrated that due to substrate effects, both the anode and the cathode should have influences to the theoretical maximum of V_{OC}. The electronic structures of the C_{60}/CuPc heterojunction are highly sensitive to (1) the ambient environment exposure, (2) the deposition sequence, and (3) substrate work function. In addition, we have demonstrated the use of rare-earth metal doping in EBL for OPV devices. Yb-doped BPhen can be a good optical spacer in terms of good conductivity and optical transmission. Besides, electron injection barrier is significantly reduced upon metal doping. Yb doping leads to a pronounced enhancement by providing electron passage while blocking the holes.

References

1. Li G, Shrotriya V, Huang J et al (2005) High-efficiency solution processable polymer photovoltaic cells by self-organization of polymer blends. Nat Mater 4:864–868
2. Kim Y, Cook S, Tuladhar SM et al (2006) A strong regioregularity effect in self-organizing conjugated polymer films and high-efficiency polythiophene:fullerene solar cells. Nat Mater 5:197–203
3. Drechsel J, Männig B, Kozlowski F et al (2005) Efficient organic solar cells based on a double p-i-n architecture using doped wide-gap transport layers. Appl Phys Lett 86:244102
4. Shrotriya V, Wu EHE, Li G et al (2006) Efficient light harvesting in multiple-device stacked structure for polymer solar cells. Appl Phys Lett 88:064104
5. Shao Y, Yang Y (2005) Efficient organic heterojunction photovoltaic cells based on triplet materials. Adv Mater 17:2841
6. Tang CW (1986) Two-layer organic photovoltaic cell. Appl Phys Lett 48:183
7. Rand BP, Burk DP, Forrest SR (2007) Offset energies at organic semiconductor heterojunctions and their influence on the open-circuit voltage of thin-film solar cells. Phys Rev B 75:115327

8. Schilinsky P, Waldauf C, Brabec CJ (2002) Recombination and loss analysis in polythiophene based bulk heterojunction photodetectors. Appl Phys Lett 81:3885
9. Karg S, Riess W, Dyakonov V et al (1993) Electrical and optical characterization of poly (phenylene-vinylene) light emitting diodes. Synth Met 54:427
10. Brabec CJ, Cravino A, Meissner D et al (2001) Origin of the Open Circuit Voltage of Plastic Solar Cells. Adv Funct Mater 11:374
11. Scharber MC, Mühlbacher D, Koppe M et al (2006) Design rules for donors in bulk-heterojunction solar cells – towards 10% energy-conversion efficiency. Adv Mater 18:789
12. Lee ST, Hou XY, Mason MG et al (1998) Energy level alignment at Alq/metal interfaces. Appl Phys Lett 72:1593
13. Ishii H, Sugiyama K, Ito E et al (1999) Energy Level Alignment and Interfacial Electronic Structures at Organic/Metal and Organic/Organic Interfaces. Adv Mater 11:605
14. Kahn A, Koch N, Gao W et al (2003) Electronic structure and electrical properties of interfaces between metals and π-conjugated molecular films. J Polym Sci B Polym Phys 41:2529
15. Vázquez H, Gao W, Flores F et al (2005) Energy level alignment at organic heterojunctions: role of the charge neutrality level. Phys Rev B 71:041306
16. Vázquez H, Flores F, Kahn A et al (2007) Induced Density of States model for weakly-interacting organic semiconductor interfaces. Org Electron 8:241
17. Lau KM, Tang JX, Sun HY et al (2006) Interfacial electronic structure of copper phthalocyanine and copper hexadecafluorophthalocyanine studied by photoemission. Appl Phys Lett 88:173513
18. Bhosle V, Prater JT, Yang F et al (2007) Gallium-doped zinc oxide films as transparent electrodes for organic solar cell applications. J Appl Phys 102:023501
19. Hong ZR, Liang CJ, Sun XY et al (2006) Characterization of organic photovoltaic devices with indium-tin-oxide anode treated by plasma in various gases. J Appl Phys 100:093711
20. Tang JX, Lau KM, Lee CS et al (2006) Substrate effects on the electronic properties of an organic/organic heterojunctions. Appl Phys Lett 88:232103
21. Tang JX, Lee CS, Lee ST (2007) Electronic structures of organic/organic heterojunctions: From vacuum level alignment to Fermi level pinning. J Appl Phys 101:064504
22. Zhou YC, Liu ZT, Tang JX et al (2009) Substrate dependence of energy level alignment at the donor–acceptor interface in organic photovoltaic devices. J Electron Spectrosc Relat Phenom 174:35–39
23. Rusu M, Strotmann J, Vogel M et al (2007) Effects of oxygen and illumination on the photovoltaic properties of organic solar cells based on phtalocyanine:fullerene bulk heterojunctions. Appl Phys Lett 90:153511
24. Song QL, Wang ML, Obbard EG et al (2006) Degradation of small-molecule organic solar cells. Appl Phys Lett 89:251118
25. Wu HR, Song QL, Wang ML et al (2007) Stable small-molecule organic solar cells with 1, 3, 5-tris(2-N-phenylbenzimidazolyl) benzene as an organic buffer. Thin Solid Films 515:8050
26. Tanaka Y, Kanai K, Ouchi Y et al (2007) Oxygen effect on the interfacial electronic structure of C60 film studied by ultraviolet photoelectron spectroscopy. Chem Phys Lett 441:63
27. Drechsel J, Männig B, Kozlowski F et al (2005) Efficient organic solar cells based on a double p-i-n architecture using doped wide-gap transport layers. Appl Phys Lett 86:244102
28. Shen L, Zhu G, Guo W et al (2008) Performance improvement of TiO2/P3HT solar cells using CuPc as a sensitizer. Appl Phys Lett 92:073307
29. Tang JX, Zhou YC, Liu ZT et al (2008) Interfacial electronic structures in an organic double-heterostructure photovoltaic cell. Appl Phys Lett 93:043512
30. Hamed A, Sun YY, Tao YK et al (1993) Effects of oxygen and illumination on the in situ conductivity of C60 thin films. Phys Rev B 47:10873
31. Ng TW, Lo MF, Zhou YC et al (2009) Ambient effects on fullerene/copper phthalocyanine photovoltaic interface. Appl Phys Lett 94:193304

32. Crispin X, Geskin V, Crispin A et al (2002) Characterization of the interface dipole at organic/metal interfaces. J Am Chem Soc 124:8131
33. Koch N, Kahn A, Ghijsen J et al (2003) Conjugated organic molecules on metal versus polymer electrodes: Demonstration of a key energy level alignment mechanism. Appl Phys Lett 82:70
34. Li N, Lassiter BE, Lunt RR et al (2009) Open circuit voltage enhancement due to reduced dark current in small molecule photovoltaic cells. Appl Phys Lett 94:023307
35. Ng TW, Lo MF, Liu ZT et al (2009) Substrate effects on the interface electronic properties of organic photovoltaic devices with an inverted C60/CuPc junction. J Appl Phys 106:114501
36. Peumans P, Bulović V, Forrest SR (2000) Efficient photon harvesting at high optical intensities in ultrathin organic double-heterostructure photovoltaic diodes. Appl Phys Lett 76:2650
37. Peumans P, Forrest SR (2001) Very-high-efficiency double-heterostructure copper phthalocyanine/C60 photovoltaic cells. Appl Phys Lett 79:126
38. Kim JY, Kim SH, Lee HH et al (2006) New architecture for high-efficiency polymer photovoltaic cells using solution-based titanium oxide as an optical spacer. Adv Mater 18:572
39. Hänsel H, Zettl H, Krausch G et al (2003) Optical and electronic contribution in double-heterojunction organic thin-film solar cell. Adv Mater 15:2056
40. Rand BP, Li J, Xue J et al (2005) Organic double-heterostructure photovoltaic cells employing thick Tris(acetylacetonato)ruthenium(III) exciton-blocking layers. Adv Mater 17:2714
41. Peumans P, Yakimov A, Forrest SR (2003) Small molecular weight organic thin-film photodetectors and solar cells. J Appl Phys 93:3693
42. Kido J, Matsumoto T (1998) Bright organic electroluminescent devices having a metal-doped electron-injecting layer. Appl Phys Lett 73:2866
43. Chan MY, Lai SL, Fung MK et al (2003) Efficient CsF/Yb/Ag cathodes for organic light-emitting devices. Appl Phys Lett 82:1784
44. Kwon S, Kim SC, Kim Y et al (2001) Photoemission spectroscopy study of Alq3 and metal mixed interfaces. Appl Phys Lett 79:4595
45. Chan MY, Lai SL, Lau KM et al (2006) Doping-induced efficiency enhancement in organic photovoltaic devices. Appl Phys Lett 89:163515

Improving Polymer Solar Cell Through Efficient Solar Energy Harvesting

Hsiang-Yu Chen, Zheng Xu, Gang Li, and Yang Yang

Contents

1 Introduction 200
2 RR-P3HT:PCBM System 201
   2.1 Slow Growth (Solvent Annealing) 202
   2.2 Mixed Solvent 212
3 Efficient Inverted Polymer Solar Cells Using Functional Interfacial Layers 220
4 Development of Low Bandgap Polymers for Polymer Solar Cells 226
5 Summary 234
References 234

Abstract In the last few years, several effective approaches have been developed to improve polymer solar cell performance. In this chapter, we summarized several of the efforts conducted in UCLA on polymer solar cells, of which each is associated to efficient light harvesting. We first discussed effective approaches to improve morphology and nanoscale structure control on the polymer active layer through (a) solvent annealing and (b) mixed solvent approaches, in order to enhance the crystallinity of polymer for enhancing absorption, charge transport, and efficiency in the RR-P3HT:PCBM system. Interface engineering work has led to the demonstration of novel inverted solar cell structure, which might have advantages over conventional device structure in solution electrode, transparent solar cell, and/or tandem structure. The third section is focused on the newly developed low bandgap polymers, which show 5.6% solar cell efficiency – a significant improvement over the model RR-P3HT:PCBM solar cell. The results are good representative of the recent progress in the field of organic solar cell.

H.-Y. Chen and Y. Yang (✉)
Department of Materials Science and Engineering, University of California, Los Angeles, CA 90095, USA
e-mail: yangy@ucla.edu

Z. Xu and G. Li (✉)
Solarmer Energy Inc, El Monte, CA 91731, USA
e-mail: gangl@solarmer.com

1 Introduction

Fossil fuels like oil and coal have been the primary energy source for human being in the past few centuries. With the world oil production probably already peaked and the world consumption still continuously increasing, the quest for renewable energy source becomes crucial for the sustainable development of human being. This is further strengthened by the heavy environmental footprint of fossil fuel, because all combusted fossil fuels emit CO_2 and contribute to global warming. Photovoltaic (PV) effect, which converts sunlight directly into electricity, is increasingly being recognized as one key technology for future world energy production. Unlike other resources, solar energy is almost unlimited. Solar irradiation on the earth is 162 kTW. In another word, 14 TW per hour, almost the same as the world's total annual energy consumption [1]. PV systems are with minimal environmental impact and they are also portable and suitable to distributed application, which could be one of the next high technology economic drivers.

Currently, silicon-based solar cell dominates the PV market with over 80% market share. The price of traditional silicon solar cell is one of the major issues limiting the widespread application of solar energy (contributes to only ~0.03% of the world energy needs) [2], with solar energy cost $0.25/kW-h vs. $0.05–0.08 for conventional energy source [3].

Organic photovoltaic cells (OPVs) have attracted intensive investigation mainly because of their potential for low cost solar energy. The low material cost, very low material consumption, very high material utilization through low cost printing or coating technologies, as well as the mechanical properties of the polymers are among the advantages of OPVs, especial polymer-based bulk heterojunction (BHJ) solar cells.

The energy conversion process in a polymer solar cell is a multiple step process [4] including the followings.

Step 1 – In-coupling of the photon: The incoming photons encounter the first interface – glass or plastics, when they hit the device. The reflection losses at the air–substrate interface as well as at each subsequent interface, which depend on the difference between optical refractive indices of the two materials, can be minimized by proper antireflection coating.

Step 2 – Photon absorption: The incoming photons are then absorbed in the active layer. The absorption spectrum of the active material should match the solar irradiation for maximum absorption. As a result, low bandgap polymers are highly desirable.

Step 3 – Exciton formation and migration: After a photon has been absorbed in the polymer, an exciton is formed. The excitons then diffuse in the material with a characteristic exciton diffusion length (L_D) typically of the order of 5–10 nm [5, 6]. L_D depends on the structure of the material and the dielectric properties. The excitons have a finite lifetime, and during the diffusion they decay or dissociate through several mechanisms. In order to achieve efficient photovoltaic conversion, the excitons have to be dissociated into free electrons and holes before they decay radiatively, thermally, or vibronically.

Step 4 – Exciton dissociation or charge separation: The most common way to achieve exciton dissociation into free electrons and holes is through photoinduced charge transfer process. The processes that act as counterforce to exciton dissociation are geminate recombination, where separated electrons and holes recombine back to form an exciton, and nongeminate bimolecular recombination, where an electron and a hole from different excitons recombine. The exciton dissociation has been shown to happen at donor/acceptor interface, where the D/A energy level difference provides the driving force for efficient exciton dissociation into charges, which can then transport selectively in donor or acceptor. Although larger D/A energy level difference can facilitate efficient exciton dissociation, the preferred energy difference should be slightly larger than exciton binding energy (by ~0.3 eV). Larger energy level difference could lead to lower solar cell efficiency since the photovoltage loss associated with it could surpass the photovoltage gain.

Step 5 – Charge transport: The free charges must then travel through the active layer to reach the electrodes where they can be collected to produce photocurrent. The charge carrier mobilities for both electron and hole, therefore, play an important role in determining device efficiency. The charge carrier mobility in conjugated polymers is usually very low, making it necessary to have a thin active layer, which on the other hand may reduce the optical absorption.

Step 6 – Charge collection: Finally, the free electrons that reach the electrodes are collected and passed into the outer circuit to generate device photocurrent. The charge collection efficiency can be affected by the energy level matching at the metal–polymer interface, interfacial defects, and so on.

In this chapter, we will start with the model solution-processable BHJ polymer solar cell system that is composed of conjugated polymer, regioregular poly(3-hexylthiopene) (RR-P3HT), and fullerene derivative, [6,6]-phenyl-C_{61}-butyric acid methyl ester ([60]PCBM). We will discuss several approaches we recently developed in UCLA to enhance the light harvesting efficiency of RR-P3HT:PCBM solar cell. The methods include solvent annealing (slow growth), mixed solvent, and new device structure. The mechanism behind the performance enhancement will be discussed by connecting to various steps described above, especially photon absorption and charge transport. Following the observation of the vertical phase separation in P3HT:PCBM system, we will introduce our recent work on the improvement of polymer solar cell based on inverted device structure. While the P3HT bandgap being 1.9 eV and the open-circuit voltage is only 0.6 V, we will introduce our recent effort in developing novel low bandgap polymers to reduce the energy loss and achieve high performance polymer solar cell.

2 RR-P3HT:PCBM System

Region-regular poly(3-hexylthiophene) was invented in the early 1990s [7–10]. RR-P3HT absorption edge is at ~650 nm matching well with the strongest solar spectrum and is about 100 nm into the red compared to former champion of MEH-PPV.

In addition, RR-P3HT has one of the best charge transport properties among conjugated polymers. The application of RR-P3HT in OPV was stimulated by the work of Padinger et al. [7], who reported 3.5% power conversion efficiency (PCE) polymer solar cells by annealing RR-P3HT:PCBM blend. The annealing conditions are then extensively studied by various groups over the World and efficiency up to 5% had been reported.

In 2005, a slow growth (solvent annealing) approach was developed by the UCLA group to enhance RR-P3HT:PCBM solar cell efficiency. Both the absorption and charge transport were shown to be improved significantly along with the balanced charge transport. These were proposed to be the main factors that contribute to the performance improvement. We here showcase our following work to illuminate the working principle further.

2.1 Slow Growth (Solvent Annealing)

In slow growth approach, the film is left in liquid phase after spin coating and kept in a confined volume (e.g., glass Petri dish) to let the solvent dry slowly. Since the slow growth method significantly improves RR-P3HT:PCBM-based solar cell efficiency as in thermal annealing approach, we use the term "solvent annealing" in this chapter. While combined with film thermal annealing, PCE over 4% can be achieved under air mass (AM) 1.5 Global (G) standard reference condition (SRC), a very impressive 3.5% PCE was achieved by solvent annealing approach alone [8]. In this section, we present a time evolution study on the effect of solvent annealing on light harvesting of the RR-P3HT:PCBM polymer solar cells with the only variable being the film spin-coating time (t_s). We further define a term called solvent annealing time, (t_a), which is the time taken by the solvent to dry after the spin-coating process. t_s and t_a are correlated: the shorter t_s is set, the longer t_a is needed, and vice versa. Spin coating itself is a fast drying process and the solvent annealing is a slow growth process. By adjusting the t_s (and thus t_a), a comprehensive investigation of the morphology evolution of RR-P3HT:PCBM polymer film and the performance of the polymer solar cells can be performed.

The experimental design was based on several facts. First, both conventional spin coating (fast drying) and solvent annealing (slow growth) transform polymer solution into a solid polymer film. Second, during spin coating, the effective solvent evaporation rate is much faster than that in an isolated glass Petri dish. Therefore, the length of t_s effectively determines the time required to dry the film, that is, the solvent annealing time t_a. Dichlorobenzene (DCB) was chosen as solvent whose high boiling point provides possible wide spin-coating time window for controllable experiment. At spin-coating speed of 1,000 rpm, solvent evaporates much faster than in a static case as in glass Petri dish. We noticed that as t_s increases from 20 to 50 s, t_a decreases dramatically from ~20 to ~1 min, judging by the drastic visual color change of the film color during the liquid (orange) to solid (dark purple) transition. As described earlier, during the spin-coating process, the majority of the solution is removed from the substrate by the centrifugal force initially and the film thickness reduces very fast. With extended time, an equilibrium state will be established and the film thickness will reach a constant. Under the specific spin speed of 1,000 rpm, Dektek profilometer results indicate that the film with $t_s = 20$ s is ~165-nm thick, and the rest of the samples ($t_s = 30-80$ s) have film thickness of 150 ± 5 nm. All the experiments were conducted at room temperature. At an elevated temperature, the solvent removing speed will increase and the blend film will lose its crystallinity. We have found that in open N_2 environment, the solvent annealing time is reduced from 3 min at room temperature to 20 s at 70°C, and the long wavelength absorption as well as device performance degrade significantly [8]. Same trend is expected when temperature is increased during spin coating, although we are limited by our experimental conditions. UV–Vis spectroscopy in Fig. 1a shows only very slight change in both magnitude and shape for solvent-annealed films with t_a between 20 min and 1 min (or the t_s between 20 and 50 s). In all the cases, the vibronic features are very clear, indicating that with merely 1 min solvent annealing, the ordering of RR-P3HT can be well maintained in 1:1 wt. ratio RR-P3HT:PCBM blend film. t_s values of 52 s and 55 s represent the transition points in film properties. The $t_s = 52$ s film was observed to just start to solidify right after the spin-coating process, and it took ~20 s to observe a full color change. When t_s increases to 55 s, the films were visually solidified right after spin coating. (However, when the films were left in glass Petri dish, a further slight color darkening was observed, indicating tiny amount of DCB solvent residue in the film.) The absorption of this film ($t_s = 55$ s) shows significant reduction in red-region absorption and the originally strong vibronic shoulders get significantly diminished. Further increasing t_s to 80 s is used to represent the extreme case of minimal solvent annealing, in which the film shows further loss of both absorption in the red region and vibronic features. For RR-P3HT:PCBM films, solvent-annealed films have the most prominent vibronic features reported in the literature, indicating strong interchain–interlayer interaction [9–12] of RR-P3HT chains as well as well-maintained polymer ordering in the blend films. Except the film with $t_s = 20$ s, the absorptions below 450 nm in all the films are very similar, indicating that all the films have similar thickness (i.e., the amount of RR-P3HT and PCBM). PCBM absorption is clearly unaffected by t_a, which indicates the driving force of device improvement is by self organization of polymer through solvent annealing process.

Conformational chain defects (twists or disruptions of planarity) can interrupt polymer conjugation and lead to smaller π-conjugation length and blue-shift of π–π* absorption band in linear conjugated polymers like P3HT [13–15]. High regioregular P3HT can form semicrystalline, self-organized lamellar morphology with superior order forms [9–12], resulting in high hole mobility in both field-effect transistor (FET) [13–15] and solar cell applications. The RR-P3HT absorption at longer wavelength as well as vibronic shoulders are reduced significantly in the spin-casting of films with significant amount of PCBM in the traditional way (no solvent annealing, yellow phase), indicating partially loss of crystallinity and/or smaller conjugation length in RR-P3HT due to the PCBM disturbance. The quick

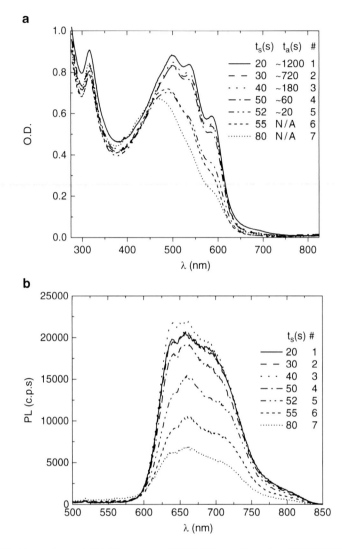

Fig. 1 (a) Absorption and (b) PL spectra of 150–165 nm thick 1:1 RR-P3HT:PCBM films fabricated by spin coating for different times (t_s from 20 s for #1 to 80 s for #7). As t_s becomes longer, the absorption in the long-wavelength region and the PL intensity are decreased. Reproduced with permission from Adv. Funct. Mater., 17, 1637, 2007. Copyright 2007, Wiley-VCH Verlag GmbH & Co. KGaA, Weinheim

solvent removing process that turns to freeze the P3HT:PCBM uniform distribution overweigh the effect of RR-P3HT interchain interaction to form highly ordered lamellae. With solvent annealing, the very preferred RR-P3HT absorption is well restored in the blend film with prominent vibronic features [16]. The absorption of the blend film, which is a simple addition of that of PCBM and RR-P3HT films, strongly supports the occurrence of phase segregation and the formation of crystalline RR-P3HT domains in the blend film.

Photoluminescence (PL) quenching is a direct indication of exciton dissociation and efficient PL quenching is a necessary condition for efficient organic solar cells. However, this does not mean the stronger the PL quenching, the better the solar cell in the same system. Figure 1b shows the PL spectra of the same set of films with different degree of solvent annealing as those in Fig. 1a. The PL spectra of films #1 to #3 with t_s of 20–40 s have similar PL intensities of 20,500–21,900 c.p.s (count per second, 280k c.p.s in pure P3HT film). With t_s of 50 s (#4), the PL intensity slightly drops to 19,200 c.p.s. Further increasing spin time, however, drastically reduces the PL intensity to 6,800 c.p.s in film #7 (t_s of 80 s), less than ~30% of those of solvent-annealed films. Because the system is at most polycrystalline in nature, the angle dependence of PL is weak. Scaling with film thickness also has minimal effect on the conclusion. Earlier study on polythiophenes shows that PL quantum efficiency of regiorandom P3HT (absorption blue shifted with respect to regioregular P3HT) has over one order of magnitude higher ($\eta = 8\%$) than that of regioregular (92% RR) P3HT ($\eta < 0.5\%$) [17]. This is explained by the fact that self-assembled RR-P3HT lamellae have strong interchain–interlayer interaction, which splits the highest occupied molecular orbitals (HOMO) and lowest unoccupied molecular orbitals (LUMO) levels, and the lowered LUMO level becomes optically forbidden [18] – that is, weaker radiative transition. This apparent controversy between absorption and PL is removed when consider the ultrafast (~40–100 fs) photoinduced charge transfer process [19] at donor/acceptor (D/A) interface, which clearly dominates the PL quantum efficiency difference. In disordered blend film, the D/A interface is larger since the D and A distribution is frozen during the spin-coating process from the homogeneous blend solution. In addition, in the local self-organized RR-P3HT domains, defect sites introduced by PCBM as well as nonradiative RR-P3HT decay channels are eliminated, which might also contribute to PL enhancement.

Eight devices were fabricated for each type of film and the device performance is shown in Fig. 2. Figure 2a shows the current–voltage (J–V) curves of the best devices of each type. Figure 2b shows the statistics data of device parameters derived from eight devices of each type. Maximum short-circuit current density (J_{SC}) is 9.6 mA cm^{-2} in device (#1) with longest t_a (~20 min, or a short t_s). However, reducing t_a to as short as 1 min ($t_s = 50$ s) only leads to a few percent J_{SC} reduction ($J_{SC} = 9.1$ mA cm^{-2}). Thicker film (#1) may slightly contribute to J_{SC}. Further increasing the spinning time leads to quick J_{SC} reduction – 4.0 mA cm^{-2} with 55 s t_s and 2.8 mA cm^{-2} with 80 s t_s. This trend – an initial plateau in J_{SC} and PCE followed with a quick drop, agrees well with both UV–Vis and PL results. Open-circuit voltage (V_{OC}) shares similar trend, with at ~0.58 V for t_s of 20–50 s; it quickly increases to 0.67 V at 80 s t_s device. Fill-factor (FF), however, does not show much change even with drastic J_{SC} variation. Except in 55 s t_s device (FF = 58.3%), all other devices have high FF ranging from 62.2 to 64.8%. We conjecture that this indicates the device deterioration due to polymer ordering loss which begins with J_{SC} reduction followed by FF lowering. With the relatively slow

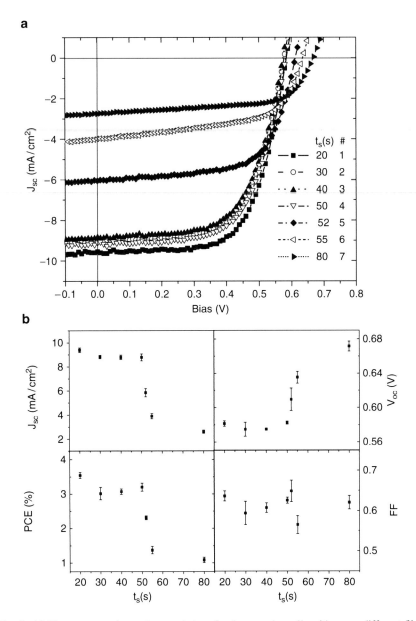

Fig. 2 (a) The current–voltage characteristics of polymer solar cells with seven different films spin coated for different times as the active layers (the films are 150–165 nm thick). (b) Device parameters derived from eight devices of each type. Reproduced with permission from Adv. Funct. Mater., 17, 1637, 2007. Copyright 2007, Wiley-VCH Verlag GmbH & Co. KGaA, Weinheim

spin-speed (1,000 rpm), even the device with 80 s spin-coating time might represent an intermediate film ordering condition, that is, a transition state from device with highly ordered film (expecting high J_{SC} and high FF) to that with highly disordered film (expecting low J_{SC} and low FF). Atomic force microscopy (AFM) morphology data and grazing incident X-ray diffraction (GIXRD) data provide further evidences (see Figs. 4 and 5). The combined effect is that PCE reduces slightly from 3.64% (20 s t_s device) to 3.38% (50 s t_s device) and then significantly drops to 1.17% in 80 s t_s device.

The ordered RR-P3HT domain is crucial for light absorption whereas the interpenetrating network of RR-P3HT and PCBM pathway is important for facilitating charge transport. This will reduce the interface area between RR-P3HT and PCBM. The high performance solar cells made from solvent-annealed films indicate that the interface area is sufficiently large (with significant PL quenching) and provides sufficient exciton dissociation. In the trade-off between (a) efficient charge separation and (b) absorption plus charge transport for maximizing polymer solar cell efficiency, the benefit of (b) clearly dominates that of (a) in the RR-P3HT:PCBM system. The size of polymer crystalline domains can also be estimated to be comparable to the exciton diffusion length (i.e., a few nm).

The V_{OC} in polymer solar cells has been widely discussed over the years. With metal–insulator–metal (MIM) [20] picture not sufficient to explain the phenomena observed, the most widely accepted belief is that in the case with ohmic contact at both electrodes, V_{OC} linearly varies with the energy difference between the donor HOMO and acceptor LUMO [6]. There are also evidences of surface dipoles in these devices [21]. In this manuscript, however, with all parameters being the same except t_s, it is interesting to see a variation of ~0.1 V in V_{OC}. Two mechanisms are conjectured to contribute to this reduction in V_{OC}. First is the formation of band structure [7] instead of molecular energy level (or alternatively the splitting of HOMO and LUMO levels [18]) due to strong interchain–interlayer interaction within relatively high ordering of RR-P3HT in solvent-annealed films, which leads to a reduced effective "band gap" relative to HOMO–LUMO difference. Second, in solvent-annealed films, the RR-P3HT has considerably longer conjugation length on average than that in films without solvent annealing [16], indicated by the much stronger intensity of absorption at longer wavelength region as discussed before. Therefore, the reduction of V_{OC} can be expected since the difference between the effective polymer "HOMO" and acceptor "LUMO" levels decreases. Chen et al. [10] had observed the band edge of the regioregular head–tail (HT) P3HT was 0.4 eV lower than that of the regiorandom P3HT. It is reasonable to relate P3HT in solvent-annealed 1:1 blend film being RR-P3HT and P3HT in nonsolvent-annealed blend film being similar to regiorandom P3HT. Similar V_{OC} changes have been reported in several papers involving thermal annealing on P3HT:PCBM systems [7, 22–26], in which thermal annealing improves crystalline ordering of RR-P3HT and, therefore, a reduction of V_{OC} can be understood.

The incident photon-to-electron conversion efficiency (IPCE) results of four devices with t_s of 20, 50, 55, and 80 s were shown in Fig. 3. Accompanied with the trend in J_{SC}, the magnitude of IPCE was first slightly reduced from 63.3%

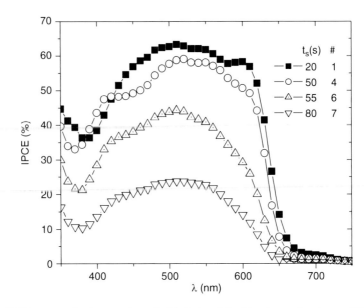

Fig. 3 IPCE spectra of polymer solar cells #1, 4, 6, and 7 with spin-coating times of 20, 50, 55, and 80 s, respectively. Both the IPCE magnitude and the photoresponse in the red region of the electromagnetic spectrum are reduced as t_s increases. Reproduced with permission from Adv. Funct. Mater., 17, 1637, 2007. Copyright 2007, Wiley-VCH Verlag GmbH & Co. KGaA, Weinheim

(#1: 20 s t_s) to 59.1% (#4: 50 s t_s), followed by a fast decrease to 44.3% in t_s of 55 s device (#6) and finally to 23.8% in t_s of 80 s (#7). In addition to this change in the magnitude, the curve shape of IPCE also undergoes clear changes. Device 1 shows a plateau in the visible region, with IPCE being over 58% from 460 to 600 nm. Device 4 starts to lose its photoresponse in the red region close to 600 nm. The IPCE of Devices 6 and 7, in which the polymer ordering was significantly lost, clearly shows a drastic monotonic reduction from their peak at 510 nm toward red region. Quantitatively, the ratios of the IPCE values at 610 nm to the peak IPCE values (i.e., IPCE ($\lambda = 610$ nm)/IPCE$_{max}$) in Devices 1, 4, 6, and 7 are 90, 83, 58, and 50%, respectively. This change in IPCE characteristics follows closely the UV–Vis trend with polymer ordering loss in the blend films. The reduced absorption in the red region leads to quick drop in IPCE in the devices.

Former studies indicate that polymer crystallization from solution into a thin film is a complex exothermic process, which could be affected by the solution evaporation rate [27]. The mesoscale film morphology in lateral direction of the RR-P3HT:PCBM (1:1 wt. ratio) films was visualized using tapping mode AFM (TM-AFM). Figure 4 shows typical TM-AFM topographic and phase images of RR-P3HT/PCBM blend films spun cast at 1,000 rpm with different t_s (30 and 80 s, respectively). It is found that the surface topography of 30s-cast film is significantly rougher than the 80s-cast film, with root-mean-square roughness of 3.0 nm (Fig. 4a) vs. 0.8 nm (Fig. 4c). The fibrillar crystalline domains of RR-P3HT are clearly visible

Improving Polymer Solar Cell Through Efficient Solar Energy Harvesting 209

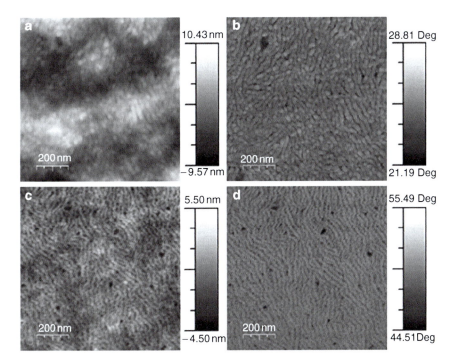

Fig. 4 TM-AFM topography (*left*) and phase (*right*) images for RR-P3HT:PCBM films fabricated using different processing conditions. The spin-coating times are (**a, b**) 30 s and (**c, d**) 80 s. Reproduced with permission from Adv. Funct. Mater., 17, 1637, 2007. Copyright 2007, Wiley-VCH Verlag GmbH & Co. KGaA, Weinheim

in both the cases, but the widths of the nanofibril crystalline domains are different. Recently, Brinkmann investigated the semicrystalline structure of RR-P3HT (MW = 35k) and found that the lamellar periodicity is approximately 28 nm [28]. This 28 nm long-range order includes the crystalline region and the disordered zones, which harbor structural defects like chain ends and folds as well as tie segments [28]. Our observation in the solar cell systems (RR-P3HT MW of 30k) agrees very well with the above findings, showing 28 nm periodicity in Figs. 4b and 4d. However, different processing condition (30 and 80 s) can result in different combination of the crystalline and disordered region in the periodicity. It was found that 80s-cast film shows longer disordered zone (crystalline grain-boundaries (GBs) even in TM-AFM topography (Fig. 4c) than that of the 30s-cast film, resulting in a high-energy barrier for interfibrillar hopping and, therefore, smaller carrier mobility [29]. This agrees to the studies reported by Kline [30] and Zhang [31] demonstrating the correlation between FET mobility and the GBs/nanofibril width of RR-P3HT in FET transistors.

Since AFM observations are limited only to the surface, the overall crystalline structure of RR-P3HT in the blend films was further studied with the aid of GIXRD analysis. As seen in Fig. 5, 2D GIXRD patterns of the 30s-cast (Fig. 5a) and 80s-cast

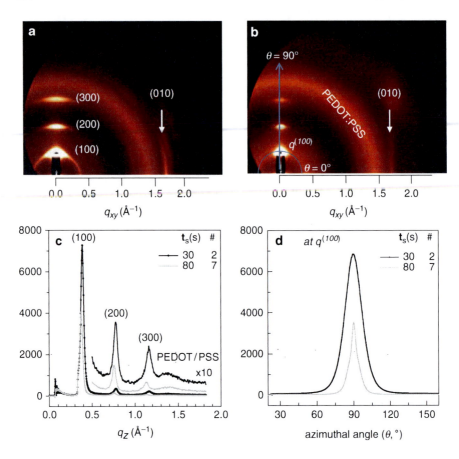

Fig. 5 2D GIXRD patterns of 1:1 RR-P3HT:PCBM films (spin coated at 1,000 rpm) fabricated with different t_s: (**a**) 30 s and (**b**) 80 s. 1D out-of-plane (**c**) X-ray and (**d**) azimuthal scan [at $q(100)$] profiles extracted from (**a**) and (**b**). The results show that the film spin coated for 30 s is more crystalline than the film spin-coated for 80 s. However, even the latter film is clearly crystalline, indicating that it is intermediate between the highly ordered and disordered phases. Longer t_a (as shorter t_s) tends to lead to a wider variety of RR-P3HT molecular orientations in the films, which is also consistent with the higher film roughness. Reproduced with permission from Adv. Funct. Mater., 17, 1637, 2007. Copyright 2007, Wiley-VCH Verlag GmbH & Co. KGaA, Weinheim

(Fig. 5b) films clearly show the intense reflections of *(100)* layer and *(010)* crystals along the q_z (substrate normal) and q_{xy} (substrate parallel) axis, respectively, implying that these films have highly ordered edge-on hexyl-side chains and parallel π-conjugated planes of RR-P3HT with respect to substrate [27, 32]. Based on 1D out-of-plane X-ray (Fig. 5c) and azimuthal angle scan (Fig. 5d) profiles extracted from the 2D GIXRD, however, we found that the 30s-cast film has much higher crystallinity than the 80s-cast film with longer t_a. In addition, from the broadness of the azimuthal peaks, longer solvent drying process possibly causes broader molecular orientation of RR-P3HT in the films. This result is consistent with relatively higher film roughness of the 30s-cast film when compared to the 80s-cast film (Fig. 4).

Based on the TM-AFM and GIXRD data, we can also conclude that the loss of P3HT long-wavelength absorption in 80s-cast film is mainly due to smaller conjugation length of the polymer rather than the degree of crystallinity. Because the 80s-cast film is an intermediate case, and FF is a parameter less sensitive to carrier mobility than photocurrent [33], it partially explains the high FF of these devices although J_{SC} reduces from 9.6 mA cm^{-2} to 2.8 mA cm^{-2}.

GIXRD results indicate that in these blend films, P3HT chains have the same edge-on orientation on the substrate as in pure P3HT film [27, 32]. For 80s-cast film, the maximum peak position of scattering vector q_z was indicated at 0.378 Å^{-1}, corresponding to the interlayer spacing of 16.6 Å, while 30s-cast (highly crystalline) film showed the interlayer spacing of 16.1 Å. The results support that the 30s-cast film has stronger interlayer interaction of RR-P3HT when compared to the 80s-cast film. Because the carrier transport is perpendicular to the polymer layers, the hole transport in devices with 30s-cast film is easier, which agrees with the J–V characteristics.

Similar results were also observed in P3HT:[70]PCBM films (1:1 in weight). Figure 6a (#1 and #2) shows 2D GIXRD patterns of P3HT:[70]PCBM films from solutions in DCB. For the sample #2, DCB is completely evaporated during spin-casting at 1,000 rpm for 90 s, while 30 s provides solvent-annealing effect on the sample #1. The 2D GIXRD patterns of the two samples imply similar highly ordered edge-on hexyl-side chains and parallel π-conjugated planes of RR-P3HT with respect to substrate as mentioned above. Moreover, Fig. 6a (#3 and #4) also shows the results of P3HT:[70]PCBM films prepared from solutions in chlorobenzene (CB). The boiling point (BP) of CB is $T_b = 131°C$ which is smaller than the BP of DCB ($T_b = 180°C$). The fast solvent evaporation conditions by relatively volatile CB induce less-ordered crystalline structure of RR-P3HT in the sample #3 (1,000 rpm, 90 s) and #4 (3,000 rpm, 90 s), as confirmed in AFM topography showing less crystalline RR-P3HT (Fig. 6d). In addition, the samples #3 and #4 contain high proportion of kinetically favorable face-on structure of RR-P3HT crystals (i.e., the side-chain parallel to the substrate), as confirmed by intense reflection of (010) crystal planes (with a layer spacing of 3.81 Å) along the q_{xy} axis in the 2D GIXRD patterns [34, 35]. From 1D out of plane X-profiles for the P3HT/[70]PCBM films examined (Fig. 6b), we found that the sample #1 with the longest crystal growth time has the highest crystallinity, although the edge-on molecular ordering becomes broader by crystal growth of RR-P3HT in a pseudo-solid-like state (Fig. 6c) An increase in polymer interlayer spacing is clearly observed – 16.3 Å in #1, 16.5 Å in #2, 16.8 Å in #3, and 16.9 Å in #4.

According to the results discussed above, the critical parameter behind the significant improvement in device performance is, therefore, believed to be the prominent properties of RR-P3HT – its high crystallinity from high regioregularity and strong interchain interaction. While the fast solvent removing process disrupts P3HT planar conformation and ordered chain packing in traditional spin-coating process, the intrinsic self-organization driving force in RR-P3HT efficiently recovers

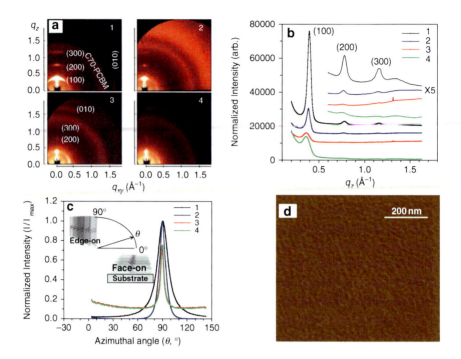

Fig. 6 (a) 2D GIXRD patterns of four RR-P3HT/[C70]PCBM (1:1 ratio) films (#1: DCB 1,000 rpm 30 s; #2: DCB 1,000 rpm 90 s; #3: CB 1,000 rpm 90 s; #4: CB 3,000 rpm 90 s). (b) 1D out-of-plane X-ray profiles. (c) Azimuthal scan profiles obtained at $q(100)$ peak positions. [The *inset* in (c) represents the correlation between the measured azimuthal angle and molecular orientation of RR-P3HT.] (d) Tapping mode AFM phase image of film #4

the destruction through a quasiequilibrium solvent annealing process. The solvent annealing process could be as short as a couple of minutes to maintain reasonably high solar cell performance.

2.2 Mixed Solvent

In addition to the solvent annealing, mixture solvent approach represents another promising method to modify solar cell morphology and improve light harvesting efficiency. Zhang et al. found a significant enhancement in photocurrent density in polyfluorene copolymer/fullerene blends when introducing a small amount of chlorobenzene into chloroform solvent [36]. Time-resolved spectroscopy on the picosecond time-scale shows that charge mobility was influenced by the mixing of solvents. Recently, Peet et al. reported that adding alkanethiol to RR-P3HT/PCBM in toluene can enhance device performance due to longer carrier lifetime with ordered structure in morphology [37]. Alkanethiol is also found effective to achieve the highest efficiency record for low bandgap polymer solar cells [38]. We conducted an in-depth investigation of the function of 1,8-octanedithiol (OT, with its chemical structure shown in Fig. 7) in the model system (RR-P3HT:PCBM) using AFM and transmission electron microscopy (TEM), X-ray photoelectron spectroscopy (XPS) combined with optical absorption and electrical measurements. RR-P3HT:PCBM with OT can phase separate into a quasioptimized morphology during spin-coating process. The mechanism of how to achieve the optimized morphology was proposed. Finally, we report two additives (shown in Fig. 7) that have similar abilities as OT to increase device performance. The model proposed here will enable us to gain insights of the mixed solvent approach in finely controlling nanomorphology in polymer solar cells.

Figure 8a shows the UV–Visible absorption spectroscopy of RR-P3HT:PCBM films of various PCBM loading ratio (33, 50, 67 wt%) processed with OT as an additive. The concentration of OT is 9.7 mg ml^{-1} in all different PCBM loadings. The films were obtained by spin coating the blend solution at 2,000 rpm for 60 s. At this spin-coating setting, all the films solidified immediately after spin coating and no further color change was observed. RR-P3HT film shows three features in absorption: two peaks at 510 and 550 nm and one shoulder at 610 nm due to strong interchain interactions [39]. When the loading of PCBM is increased, its absorption peak at 330 nm becomes more pronounced. However, it is surprising that the shape of RR-P3HT peaks remains unchanged at higher PCBM loading. The addition of OT, therefore, provides the similar function of preserving RR-P3HT crystallinity in RR-P3HT:PCBM blend as achieved in the "solvent annealing" approach. For comparison, the films processed without OT addition are shown in Fig. 8b. It is quite clear that RR-P3HT absorption peaks blue-shift at heavier PCBM loading and the originally strong vibronic shoulders diminish significantly. The effect of OT on the absorption of pure RR-P3HT (solid lines) in both Figs. 8a and 8b is rather small

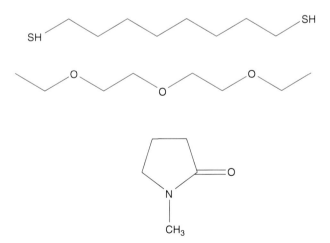

Fig. 7 Chemical structures of 1,8-octanedithiol, di(ethylene glycol)-diethyl ether, and N-methyl-2-pyrrolidinone. Reproduced with permission from Adv. Funct. Mater., 18, 1783, 2008. Copyright 2008, Wiley-VCH Verlag GmbH & Co. KGaA, Weinheim

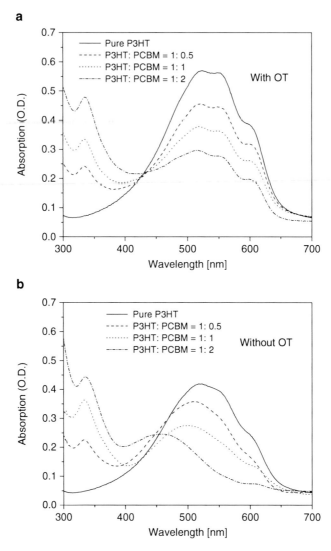

Fig. 8 Absorption of P3HT:PCBM composite layer of pure P3HT (*solid line*), P3HT:PCBM weight ratio 1:0.5 (*dashed line*), 1:1 (*dotted line*), and 1:2 (*dashed-dotted line*) processed with (**a**) and without (**b**) 1,8-octanedithiol in the blend solution. Reproduced with permission from Adv. Funct. Mater., 18, 1783, 2008. Copyright 2008, Wiley-VCH Verlag GmbH & Co. KGaA, Weinheim

because the highly regioregular P3HT has an intrinsic propensity to self-organize into microcrystalline domains [12] even without any additive. However, the difference at higher PCBM loadings (Figs. 8a and 8b) becomes more pronounced. The loss of ordering was ascribed to the fact that PCBM is finely dispersed on a molecular basis between RR-P3HT chains, thus preventing RR-P3HT from crystallizing [40]. It is, therefore, conjectured that the remarkable ability of OT is related to its capability to redistribute PCBM and RR-P3HT in the composite film.

The morphology of the P3HT:PCBM composite film was characterized using AFM. The film preparation conditions for AFM images were kept the same as those in device fabrication for accurate comparison. Figure 9 shows the typical height and phase images of blend films processed with and without OT (image size 1 × 1 mm). Two features are observed from the comparison (1) From the height images, the surface processed with OT is significantly rougher than that without OT, with root-mean-squared surface roughness of 4.0 nm (Fig. 9a) compared to 0.7 nm (Fig. 9b). Islands and valleys are apparent in Fig. 9a. It has been reported that a rough surface is a "signature" of high efficiency solar cells both in "thermal annealing" [22–26] and "solvent annealing" [8]. (2) From the phase images, highly ordered fibrillar crystalline domains of RR-P3HT are clearly visible in Fig. 9c, but

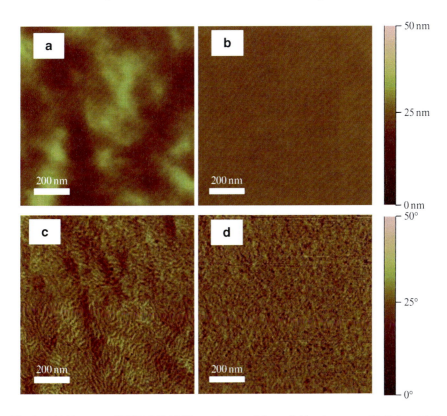

Fig. 9 AFM images of P3HT:PCBM films spin-coated from dichlorobenzene with [(**a**) and (**c**)] and without [(**b**) and (**d**)]1,8-octanedithiol. (**a**) and (**b**) are height images, and (**c**) and (**d**) are phase images. The surface processed with OT is rougher than that without OT, with root-mean-squared surface roughness 4.0 nm (**a**) compared to 0.7 nm (**b**); ordered fibrillar crystalline domains of P3HT are clearly visible in (**c**), but they are absent in (**d**). Reproduced with permission from Adv. Funct. Mater., 18, 1783, 2008. Copyright 2008, Wiley-VCH Verlag GmbH & Co. KGaA, Weinheim

they are absent in Fig. 9d. This suggests that highly ordered P3HT chain alignment is achieved when OT is added in the mixture, which is also consistent with the optical absorption measurements. While AFM is employed to probe the surface of active layer, TEM provides vertical direction information by acquisition of electrons projected through the entire film [41]. TEM has been used to distinguish RR-P3HT from PCBM due to their different densities: PCBM is 1.5 g cm^{-3} and P3HT is 1.1 g cm^{-3} [42]. We prepared our specimens for TEM measurements by spin-casting the blend solution at 2,000 rpm on glass substrate, followed by floating the film on a water surface, and transferring to TEM grids. Typical bright-field TEM images for samples processed with OT are shown in Figs. 10a and 10c. The most

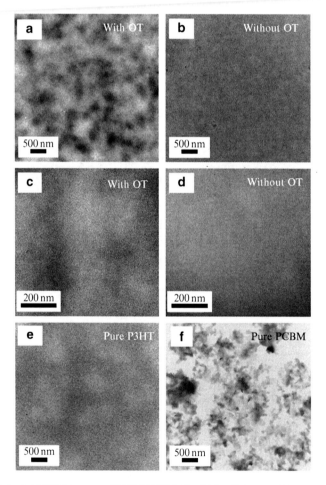

Fig. 10 Bright-field TEM images of P3HT:PCBM blend film (1:1 wt. ratio) processed with (**a**) and without (**b**) 1,8-octanedithiol additive. (**c**) and (**d**) are the zoom-in images of (**a**) and (**b**). (**e**) and (**f**) correspond to pure P3HT and PCBM samples, respectively. Reproduced with permission from Adv. Funct. Mater., 18, 1783, 2008. Copyright 2008, Wiley-VCH Verlag GmbH & Co. KGaA, Weinheim

pronounced feature compared to samples without OT (Figs. 10b and 10d) is the appearance of dark clusters in the film and high contrast of these clusters to the background. These clusters are reminiscent of TEM images by Yang et al. [42], in which PCBM-rich domains were developed during the annealing step. Similar to Yang's work, the dark regions in Fig. 10a are attributed as PCBM clusters. For comparison, pure RR-P3HT film (Fig. 10e) is very homogeneous and of low contrast, while pure PCBM (Fig. 10f) forms small crystallites with high contrast. The zoomed-in image (Fig. 10c) shows pronounced fibrillar P3HT crystals, suggesting the crystallinity has been improved compared to pristine film (Fig. 10d). To rationalize the change in morphology, we infer that the solubility of PCBM in OT plays an important role in the kinetics of the spin-coating process. The solubility of PCBM in OT and DCB measured by us show that the solubility of PCBM in OT (19 mg ml^{-1}) is much smaller than that in DCB (100 mg ml^{-1}), which also hints at the PCBM clusters shown in TEM images.

For vertical direction information, we conducted XPS measurement on the top and bottom surfaces of the active layer to determine the polymer/fullerene composition. The bottom surface was exposed by using the "floating off" method in water and collected on a TEM grid. Table 1 shows the results of peak area ratios of the S (2p) and C(1s) peaks for the top and bottom surfaces. The S(2p) peak originates from RR-P3HT and is interpreted as a signature of polymer. Even though PCBM contains oxygen, O(1s) cannot unambiguously be assigned to PCBM due to the air contamination of sample. Since C(1s) peak represents the total content of RR-P3HT and PCBM, S(2p)/C(1s) peak area ratio hence can be proportionally correlated to the RR-P3HT concentration in the blend. From Table 1 when no OT exists in the solution, we found S(2p)/C(1s) ratio in the top and bottom surfaces is very close, indicating a homogeneous distribution of RR-P3HT in the vertical direction. However, in the mixed solvent with OT, S(2p)/C(1s) ratio for the top surface is higher than that of the bottom, suggesting RR-P3HT enriched region on the surface and depleted region of the bottom. The observed vertical phase separation agrees well with other polymer/PCBM systems detected using dynamic TOF-SIMS [43] and ultraviolet photoemission spectroscopy (UPS) [44]. However, an enrichment of polymer at the bottom surface would be expected for better charge collection when poly(ethylenedioxythiophene):polystyrene sulfonate (PEDOT:PSS) coated ITO is used as substrates. The observed vertical phase separation makes it ideal for an inverted solar cell structure.

Combining all the above-mentioned results, we propose a model in the following and illustrate it in Fig. 11 to explain the above results. When RR-P3HT and PCBM are dissolved in the DCB (Fig. 11a, blue dots), polymer chains extend freely in the solvent and do not interact with PCBM. During spin coating, when DCB is

Table 1 Area ratio of S(2p) peak and C(1s) peak for the top and bottom surfaces

Peak area ratio of S(2p)/C(1s)	Top surface	Bottom surface
Without 1,8-octanedithiol	0.132	0.130
With 1,8-octanedithiol	0.140	0.106

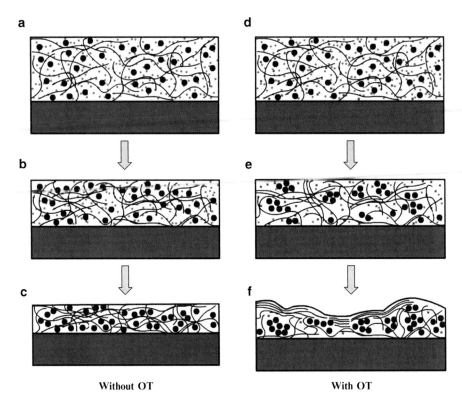

Fig. 11 Proposed model during spin-coating process. *Black wire*: P3HT polymer chain; *big black dots*: PCBM; *blue dots*: DCB molecules; and *red dots*: 1,8-octanedithiol molecules. (**a–c**) correspond to three stages in the spin-coating process when DCB is the sole solvent; (**d–f**) correspond to three stages in the spin-coating process when octanedithiol is added in DCB. Note the difference of PCBM distribution in the final stage of each case, (**c**) and (**f**). The total numbers of *big black dots* are same in all the images. Reproduced with permission from Adv. Funct. Mater., 18, 1783, 2008. Copyright 2008, Wiley-VCH Verlag GmbH & Co. KGaA, Weinheim

extracted rapidly (Fig. 11b), the whole system is quenched into a metastable state and PCBM molecules are finely dispersed between RR-P3HT chains, interrupting the ordering of P3HT chains [45] (Fig. 11c). This is supported by the blue-shifted absorption, smooth AFM morphology, and homogeneous TEM images. However, when a small amount of OT (Fig. 11d, red dots) is present in the mixed solvent, the situation is different. Because the vapor pressure of DCB is 200 times higher than OT at room temperature as shown in Table 2, DCB will evaporate much faster than OT during spin coating, and gradually the concentration of OT increases in the mixture. Due to the limited solubility of PCBM in OT (Table 2), PCBM will initially form clusters and precipitate. The fact that higher surface energy of PCBM than RR-P3HT may lead to rich PCBM distribution at the bottom of active layer when PEDOT:PSS coated ITO is used as a substrate (Fig. 11e). With a smaller

Table 2 Boiling points, vapor pressure, and PCBM solubility of DCB, OT, DEGDE, and NMP. (Boiling points and vapor pressure are from Handbook of Chemistry and Physics, 82nd Edition)

Solvent	Boiling point (°C)	Vapor pressure at 30°C (Pa)	PCBM solubility (mg ml^{-1})
1,2-dichlorobenzene	198	200	100
1,8-octanedithiol	270	1	19
di(thylene glycol)diethyl ether	189	100	0.3
N-methyl-2-pyrrolidone	229	10	18

amount of PCBM contained in the solution, RR-P3HT chains are able to self-organize in an easier fashion (Fig. 11f), supported by our former study. In this way, preformed PCBM clusters not only provide a percolation pathway for better electron transport, but also enable better hole transport in the polymer phase. As we have shown in Fig. 8, the effect of OT on the crystallization kinetics of pure P3HT is rather small compared to its effect on PCBM. In other words, the composite active layer "intelligently" phase separates into the optimum morphology in one single step rather than two stages in "thermal annealing" and takes less time than "solvent annealing." According to the proposed model, the additive in the mixture solvent approach should fulfill the following requirements: first, the compound must have lower vapor pressure than that of the primary solvent at room temperature, corresponding to a higher boiling point. Second, the compound must have lower solubility of PCBM than the solvent. Third, the compound must be miscible with solvent. Based on these requirements, we found two new additives, di(ethylene glycol)-diethyl ether (DEGDE) and N-methyl-2-pyrrolidone (NMP) that are effective in the RR-P3HT:PCBM system as well. Their chemical structures are shown in Fig. 7. I–V curves of the films processed with different mixtures under Air Mass 1.5 100 mW cm^{-2} are shown in Fig. 12. The device processed without any additive shows very limited efficiency, 0.29%, which is not surprising because no "thermal annealing" or "solvent annealing" is involved. When OT is used as an additive, J_{SC} increases to 8.14 mA cm^{-2}, FF is doubled from 31 to 63%, and PCE is almost 10 times higher. The addition of DEGDE or NMP is also found to be effective, with PCE 5–6 times higher than those without addition. DEGDE and NMP have completely different chemical structures compared to OT, suggesting that the chemical properties of OT are irrelevant to its function. However, their physical properties (PCBM solubility and room temperature vapor pressure) are similar and consistent with our hypothesis. Notably, the boiling point of the solvent used as additive cannot be too high for obtaining better performance. This is critical since such an additive will not be removed from the film during spin coating or later vacuum environment, and thermal treatment is required to remove the residue. As shown recently by Wang et al., oleic acid (boiling point 360°C) was used as a "surfactant" to enhance performance [46]; however, surfactant alone cannot improve the performance and "thermal annealing" at 155°C must be used for better performance.

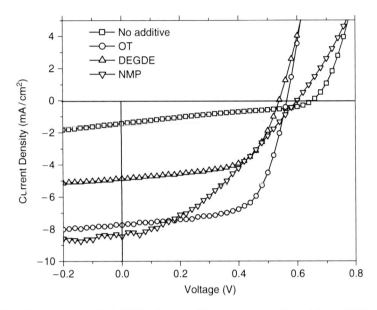

Fig. 12 *I–V* curves under AM 1.5G illumination of devices processed from (**a**) pure DCB solvent, (**b**) OT, (**c**) NMP, and (**d**) DEGDE as additive in the blend solution. Reproduced with permission from Adv. Funct. Mater., 18, 1783, 2008. Copyright 2008, Wiley-VCH Verlag GmbH & Co. KGaA, Weinheim

3 Efficient Inverted Polymer Solar Cells Using Functional Interfacial Layers

At UCLA, we have worked on inverted polymer solar cell concept. The initial motivation for the inverted configuration includes to provide design flexibility for transparent, tandem [47–51], or stacked polymer solar cells [52]. Limited absorption in the solar spectrum is the major bottleneck for high PCE, and multiple solar cells in tandem, with distinct absorption spectra, that is, different band gaps, offer the solution. Nonetheless, for solution-processed polymer solar cells, it is difficult to realize a multilayer structure without dissolution of the layers underneath. In inverted configuration the top transparent functional interfacial layer could be physically thick and provides decent protection to the underlying polymer layer against damages like (a) the solution process electrode coating in complete solution process polymer solar cell fabrication; (b) high quality transparent conducting oxides deposition through sputtering etc. for transparent solar cell or stacked cells or (c) the subsequent solution process layer deposition in tandem polymer solar cell.

As mentioned above, the vertical phase separation in the spin-coated RR-P3HT: PCBM films makes the inverted structure a favorable choice. Our recent results indicate that the PCBM concentration on some functional interfacial materials is even higher. For example, the PCBM:P3HT weight ratio at blend/Cs_2CO_3 interface can be more than 8:1, which is much higher than the 1:1 weight ratio in solution.

This property makes the inverted devices using these functional interfacial materials at cathode more attractive.

Functional interfacial material research is a very active area in the field of organic electronic devices. LiF is an effective cathode interfacial layer for both polymer-based LEDs [53] and solar cells [54]. Cesium carbonate, Cs_2CO_3, is a relatively new interfacial material, first reported for OLED applications by Canon group [55]. Unlike LiF, in an OLED the function of Cs_2CO_3 is insensitive to the metal electrode above it. Our group has demonstrated white polymer light-emitting diode (PLED) with power efficiency of 16 lm W^{-1} by using this method [56]. An enhancement in polymer solar cell efficiency was achieved by using Cs_2CO_3 interfacial buffer layer at the cathode and an efficient inverted polymer solar cell can be fabricated with the device structure: ITO/Cs_2CO_3/RR-P3HT:PCBM/V_2O_5/metal.

Conventional device fabrication processes were described in previous sections. Active layer of ~65 nm was spun coated from a 1:1 blend of RR-P3HT/PCBM solution (20 mg ml^{-1} each in dichlorobenzene) at 3,500 rpm, followed by annealing at 110°C for 10 min. V_2O_5 (10 nm), Cs_2CO_3 (1 nm), and LiF (1 nm) were thermally evaporated at the rate of about 0.02 nm s^{-1}. Solution process Cs_2CO_3 was spun coated from 2 mg ml^{-1} Cs_2CO_3 in 2-ethoxyethanol solution. The schematic of device structure is shown in Fig. 13.

In Fig. 14, the J–V curves for four different polymer solar cells with different interfacial layers at the ITO and Al interfaces are shown. In the device with no buffer layer (ITO/blend/Al), reasonable photovoltaic effect was observed with J_{SC} of 4.75 mA cm^{-2}. However, V_{OC} and FF are poor at 0.22 V and 28.5%, resulting in PCE of only 0.23%. Modifying the ITO anode by PEDOT:PSS provides significant

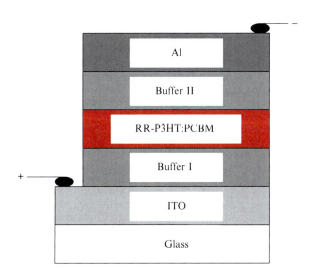

Fig. 13 Schematic of the device structure of polymer solar cell fabricated in this study. Reproduced with permission from [66]. Copyright 2006, American Institute of Physics

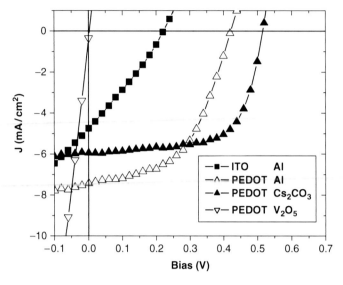

Fig. 14 The effect of interfacial layers on the performance of conventional polymer solar cells. The interfacial layers at ITO/blend and blend/Al interfaces are (**a**) none, none (*solid square*); (**b**) PEDOT:PSS, none (*open triangle*); (**c**) PEDOT:PSS, Cs_2CO_3 (*solid triangle*); and (**d**) PEDOT:PSS, V_2O_5 (*open inverted triangle*). The light intensity is 130 mW cm^{-2}. Reproduced with permission from [66]. Copyright 2006, American Institute of Physics

improvement in the device performance where J_{SC} increases to 7.44 mA cm^{-2}, V_{OC} to 0.42 V, FF to 51.8%, and the overall efficiency to 1.25%. Furthermore, the insertion of 1 nm thermally evaporated Cs_2CO_3 layer at the polymer/Al interface leads to a reduction in J_{SC} at 5.95 mA cm^{-2}. However, the V_{OC} increases to 0.52 V, and an excellent FF of 65.6% is achieved. This results in a PCE of 1.55% (~25% improvement). These results clearly show that Cs_2CO_3 can act as a functional interfacial layer to enhance polymer solar cell efficiency. The work function of PEDOT:PSS is 5.0 eV, which is 0.3 eV higher than that of ITO (4.7 eV). This work-function increase can explain the increase in V_{OC} by 0.2 eV according to the MIM model [20]. This apparently contradicts the common belief that the energy level difference between the donor HOMO and the acceptor LUMO levels dominates the V_{OC} in the polymer BHJ solar cell [6]. However, the contact changes from non-Ohmic in the case of ITO to Ohmic for PEDOT:PSS, and both electrodes being Ohmic [57] is a necessary condition for the above belief to be valid. Previous study on Cs_2CO_3 indicated that during thermal evaporation, Cs_2CO_3 decomposes into cesium oxide. Depending on the film thickness, the resulting cesium oxide has a field-emission work function of ~1.1 eV [58] because of thermoionic emission. UPS measurements conducted in our laboratory on thermally evaporated Cs_2CO_3 films show a work function of 2.2 eV. The polymer/Cs_2CO_3 contact is, therefore, Ohmic. An increase in V_{OC} by only 0.1 V, despite the work-function difference between Cs_2CO_3 and Al of 2 eV, agrees well with the earlier observation by Brabec et al. [6] and indicates Fermi-level pinning. The ITO/V_2O_5/blend/Al device shows the same polarity as that of ITO/PEDOT:PSS/polymer blend/Al device. The V_{OC} for the former is 0.38 eV, also significantly higher compared to bare ITO electrode. The HOMO level of a thermally evaporated V_2O_5 film was determined by UPS to be 4.7 eV, which is identical to that of ITO. The most plausible reason for V_{OC} enhancement is the formation of surface dipoles between V_2O_5 and an active layer, which causes an upward shift in work function of at least 0.2 eV. These results indicate that V_2O_5 can be considered as an effective hole injection layer, like PEDOT:PSS, with a similar effective work function. This is further evidenced by the J–V curves for an ITO/PEDOT:PSS/polymer blend/V_2O_5/Al device, where no photovoltaic effect was observed (see Fig. 14). The anode contact, therefore, has an important effect on polymer solar cell performance.

Figure 15 shows J–V curves for various inverted polymer solar cell structures. The ITO/polymer blend/V_2O_5 (10 nm)/Al inverted solar cell has J_{SC} = 6.97 mA cm^{-2}, V_{OC} = 0.30 V, FF = 41.2%, and PCE of 0.66%. This provides further evidence for the presence of surface dipoles that enhance the V_2O_5 work function by ~0.3 eV. Based on the results so far, an efficient inverted polymer solar cell can be achieved with the structure ITO/Cs_2CO_3/polymer blend/V_2O_5/Al, where Cs_2CO_3 was either thermally evaporated or spun coated. The J–V curves in Fig. 15 for solar cells with thermally evaporated (1 nm, open circle) and solution processed (solid triangle) Cs_2CO_3 clearly demonstrate efficient inverted solar cells. The J_{SC}, V_{OC}, and FF are very similar for the evaporated (8.42 mA cm^{-2}, 0.56 V, and 62.1%) and the solution processed (8.78 mA cm^{-2}, 0.55 V, and 56.3%) device. The overall efficiencies are 2.25 and 2.10%, respectively. Therefore, inserting V_2O_5 and Cs_2CO_3 interfacial layers can result in efficient conventional as well as inverted polymer solar cells. As a commonly used electron injection layer, ~1 nm thick LiF with ~10 nm thick V_2O_5 as hole injection layer, were also used to fabricate inverted solar cells. This device has a current density comparable to the device with Cs_2CO_3 and V_2O_5, but an antidiode behavior results in low V_{OC} (0.39 V), FF (40.7%), and PCE (0.99%).

In conventional device structure, introducing 1 nm thermally evaporated Cs_2CO_3 reduces the device photocurrent but improves V_{OC} and FF significantly, indicating possible physical damage. However, in the inverted structure, where Cs_2CO_3 is deposited on ITO substrates, all three parameters are improved. The improvements may due to the effect of the vertical phase separation.

Based on information collected, we can treat V_2O_5 as a hole injection layer with "effective" work function of ~5.0 eV and Cs_2CO_3 as an electron injection layer with very low work function. The polarity of the device is decided by the relative positions of these two interfacial layers and is insensitive to the conducting electrodes. The energy level diagrams for various inverted configurations are illustrated in Fig. 15b.

Our recent study indicates that further improvement of the efficiency of the inverted solar cell can be achieved by a thermal annealing treatment of the solution processed Cs_2CO_3 interfacial layer. Various annealing temperatures of Cs_2CO_3 interfacial layer were carried out on the hot plate inside the glove box for 20 min. Figure 16a shows J–V curves of devices with different Cs_2CO_3 annealing

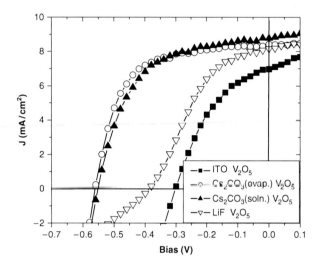

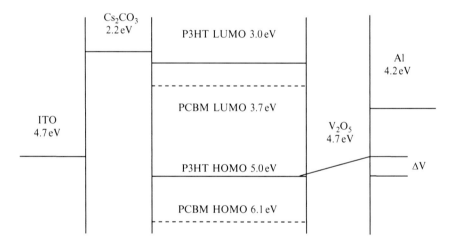

Fig. 15 (a) Current–voltage characteristics of inverted polymer solar cells. The interfacial layers at ITO/blend and blend/Al interfaces are (1) none, V_2O_5 (*solid square*); (2) Cs_2CO_3 (evaporated), V_2O_5 (*open circle*); (3) Cs_2CO_3 (solution process), V_2O_5 (*solid triangle*); and (4) LiF, V_2O_5 (*open inverted triangle*). (b) Energy level diagrams for various materials in the inverted solar cells. Reproduced with permission from [66]. Copyright 2006, American Institute of Physics

temperatures. The structure of the devices is shown in Fig. 16b, and the P3HT: PCBM active layers in these devices were fabricated using slow growth method (spun coated at 600 rpm for 40 s) and 110°C annealing for 10 min. For the device without thermal annealing on Cs_2CO_3 layer, the PCE is 2.31%. When Cs_2CO_3 layers were treated by different temperature annealing process, all device performance are improved. As the annealing temperature of Cs_2CO_3 layer increases from room temperature to 150°C, the PCE increases from 2.31 to 4.19%. In addition, all

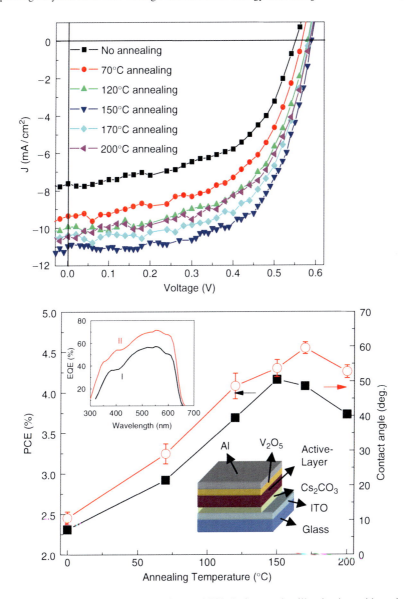

Fig. 16 (**a**) *I–V* characteristics of the inverted PV devices under illumination with various annealing temperatures of the Cs_2CO_3 layer (**b**) Power conversion efficiency (PCE) and contact angle with water of the Cs_2CO_3 layer as a function of different annealing temperatures. The *inset* in (**b**) shows the effect of annealing treatment on external quantum efficiency. *Line I* is Cs_2CO_3 layer without annealing, and *line II* is after 150°C annealing. Reproduced with permission from [63]. Copyright 2006, American Institute of Physics

Table 3 J_{SC}, V_{OC}, PCE, and FF of various inverted PV devices, and contact angle of Cs_2CO_3 layer with water upon different thermal annealing temperatures

Anneal Temperature (°C)	V_{OC} (V)	J_{SC} (mA cm^{-2})	PCE (%)	FF (%)	Contact angle (°)
0	0.55	7.61	2.31	55.2	11
70	0.565	9.36	2.92	55.2	29
120	0.58	10.86	3.69	58.6	49
150	0.59	11.13	4.19	64.0	54
170	0.582	11.29	4.08	62.1	60
200	0.584	11.39	3.73	56.0	53

J_{SC} short-circuit current density, V_{OC} open-circuit voltage, PCE power conversion efficiency, FF fill factor

other device characteristics, such as V_{OC}, J_{SC}, and FF, improved as well. Device operation parameters are summarized in Table 3. The optimum annealing temperature is determined within the range of 150–170°C. In this annealing temperature range, the average device PCE is approximately 4%, and the highest PCE achieved is 4.2% for 150°C annealing.

UPS results show that annealing lowers the work function of the Cs_2CO_3 layer. Further XPS results reveal that Cs_2CO_3 can decompose into low work function doped cesium oxide Cs_2O upon annealing. The doped cesium oxide behaves as a n-type semiconductor, with a lower interface resistance than pristine Cs_2CO_3, as well as having a relatively low work function. Benefited from the improved Cs_2CO_3 interfacial layer and the vertical phase separation in RR-P3HT:PCBM system, the inverted device shows higher external quantum efficiency (EQE) maximum (72% compared to 63%) and J_{SC} (11.13 vs. 10.6 mA cm^{-2}) in comparison to our best regular configuration device based on RR-P3HT:PCBM blends. This highly efficient inverted cell can be applied to design a multiple-device stacked polymer solar cell or a tandem cell, which are widely accepted to further improve the efficiency of polymer solar cells.

4 Development of Low Bandgap Polymers for Polymer Solar Cells

The two most decisive parameters regarding polymer solar cell efficiencies are the V_{OC} and the J_{SC}. J_{SC} is mostly determined by the light absorption ability of the material, the charge separation efficiency, as well as the high and balanced carrier mobilities. On the other hand, V_{OC} is limited by the difference in the HOMO of the donor and the LUMO of the acceptor, where a small V_{OC} (as compared to the photon energy) represents a smaller driving force for the PV process. For the RR-P3HT:PCBM system, the V_{OC} is around 0.6 V, which significantly limits the overall device efficiency. An effective method to improve the V_{OC} of polymer solar cells is to manipulate the HOMO level of the donor and/or LUMO level of the acceptor [6]. Until now, fullerene derivatives have proved to be one of the best and most commonly used electron acceptors. Fortunately, it is convenient to change the bandgap and energy levels of the donor material by modifying the chemical structure to achieve a high V_{OC} [6, 59].

Amongst various polymers, (poly[2,7-(9-(20-ethylhexyl)-9-hexyl-fluorene)-alt-5,50-(40,70-di-2-thienyl-20,10,30-benzothiadiazole]) (PFDTBT) has a deep HOMO level, which leads to a large V_{OC} when blended with PCBM. Svensson et al. [60] had reported polymer photovoltaic cells with a V_{OC} of 1 V based on alternating copolymer PFDTBT blended with PCBM. Moreover, Inganäs et al. [61] reported a systematic study of photovoltaic cells using four different fluorene copolymers by varying the length of the alkyl side chain and chemical structure, exhibiting power conversion efficiencies above 2–3%. Unfortunately, in their cases, the low photocurrent becomes a major limiting factor in achieving higher efficiencies, suggesting low carrier mobilities in the polymer blend.

Recently, poly{[2,7-(9,9-bis-(2-ethylhexyl)-fluorene)]-*alt*-[5,5-(4,7-di-2′-thienyl-2,1,3-benzothiadiazole)]} (BisEH-PFDTBT) and poly{[2,7-(9,9-bis-(3,7-dimethyl-octyl)-fluorene)]-*alt*-[5,5-(4,7-di-2′-thienyl-2,1,3-benzothiadiazole)]} (BisDMO-PFDTBT), which have the same polymer backbone as PFDTBT but with different side chains (Fig. 17a, b), were studied [62] in order to achieve higher efficiency values, as well as to investigate the side chain effects. BisDMO-PFDTBT has proven to be a promising candidate as a donor material for high efficiency polymer BHJ solar cells. Under simulated solar illumination of AM 1.5G (100 mW cm^{-2}), the BisDMO-PFDTBT blended with (6,6)-phenyl-C_{71}-butyric acid methyl ester ($PC_{71}BM$) achieved a maximum PCE of up to 4.5% with a thin active layer thickness of only 47 nm. The device exhibited an V_{OC} of 1 V, a J_{SC} of 9.1 mA cm^{-2}, and a reasonably high EQE exceeding 50% over the entire visible range, with an EQE maximum of 67% at 380 nm.

While the sun has over 60% of its photon flux in the range of 600–1,000 nm, using low bandgap polymers for solar cells [63–70] to absorb more infrared light has become a major strategy to improve the cell efficiency. It has been estimated [67] that polymers having a bandgap smaller than 1.74 eV and a carrier mobility higher than 10^{-3} cm^2 V^{-1} s^{-1} may provide a route for PCE exceeding 10%. Many

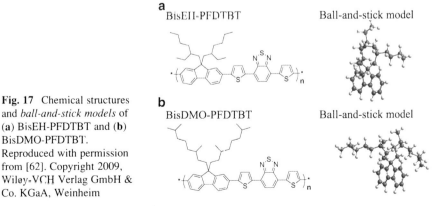

Fig. 17 Chemical structures and *ball-and-stick models* of (**a**) BisEH-PFDTBT and (**b**) BisDMO-PFDTBT. Reproduced with permission from [62]. Copyright 2009, Wiley-VCH Verlag GmbH & Co. KGaA, Weinheim

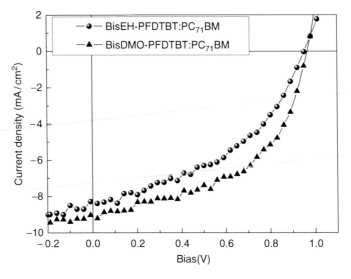

Fig. 18 The current density–voltage (*J–V*) characteristics of photovoltaic devices based on BisEH-PFDTBT:$PC_{71}BM$ (1:3, w/w) and BisDMO-PFDTBT:$PC_{71}BM$ (1:3, w/w) polymer blends, under AM 1.5G simulated solar illumination (100 mW cm^{-2}). Reproduced with permission from [62]. Copyright 2009, Wiley-VCH Verlag GmbH & Co. KGaA, Weinheim

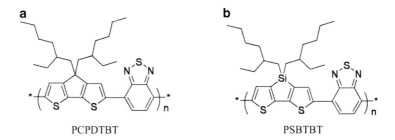

Fig. 19 Structures of (**a**) PCPDTBT and (**b**) PSBTBT. By replacing the carbon 5-position of PCPDTBT with a Si atom, a silole-containing polymer (PSBTBT) is obtained

efforts have been done in order to obtain high efficiency solar cells based on low bandgap polymers. However, so far most of the solar cells based on low bandgap polymers show efficiencies lower than 3.5% [63, 68–72]. Some reasons for the low efficiency are attributed to a low carrier mobilities, low V_{OC} [72], or inefficient carrier dissociation [68]. Recently, an efficient solar cell based on a low bandgap polymer, poly[(4,4-bis(2-ethylhexyl)-cyclopenta-[2,1-*b*;3,4-*b*′] dithiophene)-2,6-diyl-*alt*-(2,1,3-benzothiadiazole)-4,7-diyl], (PCPDTBT, Fig. 19a), was reported by using additives to control the desired morphology [38]. By adding additives such as OT or 1,8-diiodooctane into the PCPDTBT/PCBM solutions before spin coating, the morphology of the polymer film can be changed, which benefits the photovoltaic conversion processes. Even though PCPDTBT shows promising PCE results, it was mentioned that the batch-to-batch variation of PCPDTBT is large [73], possibly due to the complex synthesis process involved.

In order to ensure effective charge carrier transport to the electrodes and to reduce the photocurrent loss in solar cells, high carrier mobility is needed for polymer solar cells [8]. Poly[(9,9-dialkyl-fluorene)-2,7-diyl-*alt*-(4, 7-bis(2-thienyl)-2,1,3-benzothiadiazole)-5,5′-diyl] (PFDTBT) and its derivatives are a well-known class of photovoltaic materials, and the efficiency of this kind of material is around 2–3% [60, 74]. When the carbon atoms on the ninth position of the fluorene units of PFDTBT were substituted with silicon atoms, the photovoltaic properties of the device improved significantly, and a high PCE of 5.4% was reported by using poly [(2,7-dioctylsilafluorene)-2,7-diyl-*alt*-(4,7-bis(2-thienyl)-2,1,3-benzothiadiazole)-5,5′-diyl], PSiFDTBT, as the active layer material [59]. As reported, although the absorption spectrum of PSiFDTBT is similar to that of PFDTBT, the substitution of silicon atoms significantly improves the hole transport property of the material. As a result, the FF and J_{SC} of the PSiFDBT-based device is better than the PFDTBT-based device [59].

While silole-containing polymers have been known to show high carrier mobilities [75], attempt to utilize the advantages of silole-containing polymers and the small bandgap of PCPDTBT was thus generated. In 2008, we developed an easy and high yield synthesis method [76] for a silole-containing polymer, poly[(4,4′-bis(2-ethylhexyl)dithieno[3,2-*b*;2′,3′-*d*]silole)-2,6-diyl-*alt*-(2,1,3-benzothidiazole)-4,7-diyl], (PSBTBT, Fig. 19b). The structure of PSBTBT was kept the same as PCPDTBT in order to preserve the small bandgap. On the other hand, to transform PCPDTBT to a silole-containing polymer, the carbon 5-position is replaced with a silicon atom. The HOMO and LUMO levels of PSBTBT were measured by electrochemical cyclic voltammetry (CV). As shown in Fig. 20a, both the n- and p-doping processes of this polymer are reversible. The onset points of the n- and p-doping processes are 0.25 and –1.52 V, and the HOMO level and LUMO level of the polymer were calculated to be –5.05 and –3.27 eV, respectively. Absorption spectra

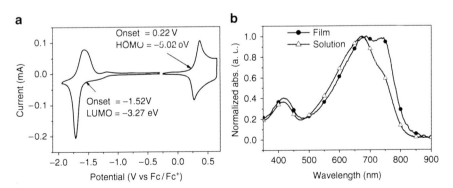

Fig. 20 (a) Cyclic voltammograms of PSBTBT films on a platinum electrode in 0.1 mol/L Bu$_4$NPF$_6$, CH$_2$CN solution; (b) absorption spectra of PSBTBT solution in chloroform and as a film. Reproduced with permission from [76] Copyright 2008, American Chemical Society

of the polymer PSBTBT in solution and in solid state are shown in Fig. 20b. The optical bandgap is 1.45 eV, which is very similar to the polymer PCPDTBT [69].

Polymer solar cells based on this PSBTBT:PC$_{70}$BM blend represent high intrinsic efficiency (~3.8%), without any additives or post treatments such as thermal annealing or solvent annealing. The optimized weight ratio of PSBTBT: PC$_{70}$BM is found to be 1:1 in chlorobenzene (CB) and 1:1.5 in chloroform. The solutions are spun coated on ITO (indium tin oxide)/glass substrates with a precoated PEDOT: PSS layer. The PSBTBT:PC$_{70}$BM films are spun coated at 5,000 rpm and the thicknesses before and after annealing are ~100 and 85 nm, respectively. A bilayer electrode composed of calcium (Ca, 20 nm) and aluminum (Al, 100 nm) is then thermally evaporated on top of the polymer films. A J_{SC} as high as 10.9 mA cm^{-2} can be obtained without any posttreatment. After annealing the spun-coated films of PSBTBT:PC$_{70}$BM at 140°C for 5 min, more than 45 devices with efficiency higher than 5% (average = 5.3%) were obtained at different times; the highest efficiency achieved was 5.6% after correcting with the mismatch factor [77]. The high J_{SC} obtained from devices with low PC$_{70}$BM loading without any posttreatment suggests a superior carrier collection efficiency, either from more effective exciton dissociation or better carrier transportation of the PSBTBT:PC$_{70}$BM system.

For comparison, PCPDTBT is synthesized by the method reported [69] and devices based on PCPDTBT:PC$_{70}$BM (1:3 wt. ratio) show comparable PCE to the reported value (~3.1%, without any additives) [38, 69]. Three different types of devices based on PCPDTBT and PSBTBT are made and the $J–V$ curves of these three devices are shown in Fig. 21 with their parameters listed in Table 4. As shown in Fig. 21, the PSBTBT-based device (black open squares) represents a better performance with a higher V_{OC} and FF than the device based on PCPDTBT (blue circles). After thermal annealing, both the J_{SC} and the FF of the PSBTBT device increased significantly while V_{OC} only dropped 0.02 V (from 0.68 to 0.66 V, red open circles). As a result, the PCE increased from 3.8 to 5.6% after the thermal treatment. It was reported that thermal annealing does not improve the device performance of solar cells based on PCPDTBT [38]. Similarly, there is no observation of efficiency improvement after thermal annealing in their PCPDTBT-based devices. The discernible response to thermal annealing after replacing the carbon 5-position of PCPDTBT with a Si atom is mainly related to the different self-assembly characteristics of these two polymers. While silole-containing organic materials are predicted to have a smaller bandgap compared with their carbon counterpart [75], a slightly larger bandgap (~1.5 eV, estimated from the absorption edge) is observed for PSBTBT compared with that of PCPDTBT (~1.46 eV) [38]. The absorption edge of spun-coated PSBTBT film is ~820 nm, as shown in Fig. 22 (green squares). Blending PC$_{70}$BM significantly increases the absorption in the visible range (red line) and no change in absorption is observed after annealing the PSBTBT:PC$_{70}$BM film (black open circles). Significant red shift of the absorption edge of PSBTBT in solution (0.001% in CB, blue line) to the solid state (spun-coated film) is observed, suggesting its strong intermolecular interactions in the solid state [39]. The absorption peak of PSBTBT film at ~740 nm is attributed to the

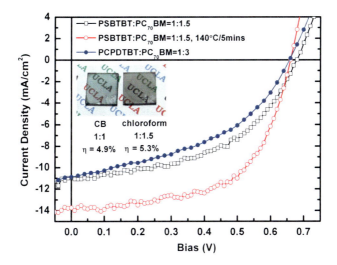

Fig. 21 Current density vs. voltage (J–V) curves of three different devices. J–V curves of PCPDTBT:PC$_{70}$BM (1:3 wt. ratio) and PSBTBT:PC$_{70}$BM (1:1.5 wt. ratio) devices fabricated under optimized conditions are compared. The PSBTBT:PC$_{70}$BM device without any posttreatment has higher V_{OC} and FF when compared with the PCPDTBT:PC$_{70}$BM device. A significant improvement in device performance is observed after annealing PSBTBT:PC$_{70}$BM at 140°C for 5 min. *Inset*: Films processed with chlorobenzene (CB) and chloroform under optimized conditions showing high transparency. Reproduced with permission from *Adv. Mater.*, 22, 371, 2010. Supporting Information. Copyright 2010, Wiley-VCH Verlag GmbH & Co. KGaA, Weinheim

Table 4 Solar cell parameters of three different devices

	V_{OC} (V)	J_{SC} (mA cm^{-2})	FF (%)	PCE (%)	Solvent
PCPDTBT:PC$_{70}$BM (1:3 wt. ratio)	0.66	−10.8	43.4	3.1	Chlorobenzene
PSBTBT:PC$_{70}$BM (1:1.5 wt. ratio)	0.68	−10.9	50.9	3.8	Chloroform
PSBTBT:PC$_{70}$BM (1:1.5 wt. ratio) Annealed 140°C/5 min	0.66	−13.6	62.2	5.6	Chloroform

strong π–π interaction of PSBTBT molecules. As a result, the intensity of the peak at 740 nm slightly decreases with the presence of PC$_{70}$BM.

The EQE spectra of three types of devices are shown in Fig. 23. Both of the EQE curves for the PSBTBT:PC$_{70}$BM devices show even distribution within the range of 350–800 nm with an EQE higher than 40%. Compared with the device without thermal annealing (black open squares), the EQE curve of the thermally annealed PSBTBT:PC$_{70}$BM device (red open circles) shows a significant increase in the range of 400–800 nm. This increment in EQE likely arises from more efficient exciton dissociation or from better carrier transportation since no obvious absorption increase is observed (Fig. 22). The shift of the PC$_{70}$BM peak is likely due to the optical effect [78] from the thickness change after thermal annealing (thickness is

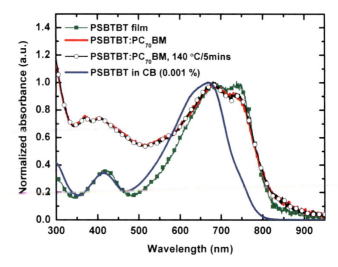

Fig. 22 Absorption spectrum of PSBTBT and PSBTBT/PC$_{70}$BM in different states. The large red-shift of the absorption edge of pure PSBTBT from solution to solid film suggests strong interchain interaction in solid state. Reproduced with permission from *Adv. Mater.*, 22, 371, 2010. Copyright 2010, Wiley-VCH Verlag GmbH & Co. KGaA, Weinheim

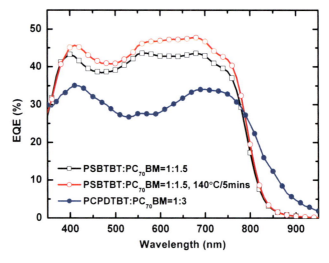

Fig. 23 External quantum efficiency of devices based on PSBTBT and PCPDTBT

slightly decreased after thermal annealing, from ~100 to 85 nm). Note that the EQE curve of the device based on PCPDTBT:PC$_{70}$BM (blue circles) starts from a wavelength longer than 900 nm and increases gradually while devices based on PSBTBT:PC$_{70}$BM show a sharp rise at the absorption edge (~820 nm). The PL of a PSBTBT film with (red line) and without PC$_{70}$BM (black line) are compared to investigate the PL quenching effect of the pure polymer after blending with

PC$_{70}$BM. The conditions for making the PSBTBT:PC$_{70}$BM film are the same as that used for fabricating devices (both with and without annealing).

Thin-film transistors (TFT) are made with PSBTBT to measure the change in hole mobility before and after annealing. The mobility (extracted from the saturation regime, see Fig. 24) obtained after annealing at 140°C for 5 min is ~3 × 10^{-3} cm^2

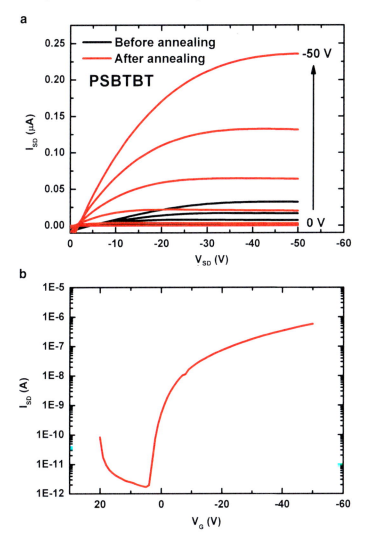

Fig. 24 Device characteristics of a thin-film transistor (TFT) made with PSBTBT. (**a**) TFT response plots of devices fabricated with PSBTBT before (*black*) and after (*red*) annealing. The applied gate voltages (V_G) are from 0 to −50 V with −10 V intervals. A significant improvement in device performance is observed after thermal annealing the device at 140°C for 5 min. (**b**) Transfer plot of the device made with PSBTBT (after annealing) at $V_{SD} = -50$ V. Reproduced with permission from Adv. Mater., 22, 371, 2010. Supporting Information. Copyright 2010, Wiley-VCH Verlag GmbH & Co, KGaA, Weinheim

$V^{-1} S^{-1}$ which is about five times higher than that before annealing (~6×10^{-4} cm^2 V^{-1} S^{-1}). A mobility higher than 10^{-3} cm^2 V^{-1} S^{-1} is desired for high efficiency polymer solar cells in order to have efficient charge transportation [67]. The increase in mobility after thermal annealing is likely to be the reason for the increased J_{SC} and FF (Fig. 21 and Table 4).

5 Summary

In summary, polymer solar cell is a promising technology. Several important approaches have been developed in the last half decade to improve polymer solar cell performance, in which all can be associated to efficient light harvesting. The device engineering side of the approaches include morphology and structure control in nanoscale on the polymer active layer, the modification of the interfaces, and novel structures like inverted and tandem structure. In this chapter, we discussed several of our works on using (a) solvent annealing and (b) mixed solvent approaches to enhance crystallinity of polymer for enhancing absorption, charge transport, and efficiency. The success of inverted structure solar cell has been linked to the morphology results of vertical phase separation, which represents an example of mutual support/benefits between practical device engineering and mechanism study. The preliminary results of polymer solar cell based on newly developed low bandgap polymers show 5.6% solar cell efficiency, a significant improvement over the model RR-P3HT:PCBM solar cell. With these promising results obtained from low bandgap polymers, higher efficiency polymer solar cells can be expected by increasing the V_{OC} without sacrificing J_{SC} and FF. Further studies on morphology control through molecular structure design, solvent mixture, and other methods will be the major challenges in order to obtain high efficiency polymer solar cells.

Acknowledgments We appreciate the financial support from the Office of Naval Research (Grant No. N00014-04-1-0434) and the Air Force Office of Scientific Research (FA9550-07-1-0264). Solarmer Energy Inc. provided part of the research funding.

References

1. Crabtree GW, Lewis NS (2007) Physics Today 60:37
2. Rühl C (2007) Statistical review of world energy. BP
3. US Department of Energy, Solar energy technologies program: solar America initiative, energy efficiency and renewable energy (EERP), Washington DC (2007)
4. Persson NK, Inganas O (2005) Simulations of optical processes in organic photovoltaic devices. In: Sun SS, Sariciftci NS (eds) Organic photovoltaics: mechanisms, materials, and devices. CRC, Boca Raton, FL
5. Peumans P, Yakimov A, Forrest SR (2004) J Appl Phys 95:2938
6. Brabec CJ, Cravino A, Meissner D et al (2001) Adv Funct Mater 11:374
7. Padinger F, Rittberger RS, Sariciftci NS (2003) Adv Funct Mater 13:85
8. Li G, Shrotriya V, Huang J et al (2005) Nat Mater 4:864

9. McCullough RD, Tristramnagle S, Williams SP et al (1003) J Am Chem Soc 115:4910
10. Chen TA, Wu XM, Rieke RD (1995) J Am Chem Soc 117:233
11. Sirringhaus H, Tessler N, Friend RH (1998) Science 280:1741
12. Sirringhaus H, Brown PJ, Friend RH et al (1999) Nature 401:685
13. Inganas O, Salaneck WR, Osterholm JE et al (1988) Synth Met 22:395
14. Hotta S, Rughooputh S, Heeger AJ (1987) Synth Met 22:79
15. Lim KC, Fincher CR, Heeger AJ (1983) Phys Rev Lett 50:1934
16. Sundberg M, Inganas O, Stafstrom S et al (1989) Solid State Commun 71:435
17. Korovyanko OJ, Osterbacka R, Jiang XM et al (2001) Phys Rev B 64:235122
18. Cornil J, Beljonne D, Calbert JP et al (2001) Adv Mater 13:1053
19. Sariciftci NS, Smilowitz L, Heeger AJ et al (1992) Science 258:1474
20. Parker ID (1994) J Appl Phys 75:1656
21. Veenstra SC, Heeres A, Hadziioannou G et al (2002) Appl Phys Mater Sci Process 75:661
22. Li G, Shrotriya V, Yao Y et al (2005) J Appl Phys 98:043704
23. Ma WL, Yang CY, Gong X et al (2005) Adv Funct Mater 15:1617
24. Reyes-Reyes M, Kim K, Carroll DL (2005) Appl Phys Lett 87:083506
25. Reyes-Reyes M, Kim K, Dewald J et al (2005) Org Lett 7:5749
26. Kim Y, Cook S, Tuladhar SM et al (2006) Nat Mater 5:197
27. Yang HC, Shin TJ, Yang L et al (2005) Adv Funct Mater 15:671
28. Brinkmann M, Wittmann JC (2006) Adv Mater 18:860
29. Shimomura T, Sato H, Furusawa H et al (1994) Phys Rev Lett 72:2073
30. Kline RJ, McGehee MD, Kadnikova EN et al (2003) Adv Mater 15:1519
31. Zhang R, Li B, Iovu MC et al (2006) J Am Chem Soc 128:3480
32. Chang JF, Sun BQ, Breiby DW et al (2004) Chem Mater 16:4772
33. Mihailetchi VD, Wildeman J, Blom PWM (2005) Phys Rev Lett 94:126602
34. Mihailetchi VD, Koster LJA, Blom PWM et al (2005) Adv Funct Mater 15:795
35. Yang H, LeFevre SW, Ryu CY et al (2007) Appl Phys Lett 90:172116
36. Zhang FL, Jespersen KG, Bjorstrom C et al (2006) Adv Funct Mater 16:667
37. Peet J, Soci C, Coffin RC et al (2006) Appl Phys Lett 89:212505
38. Peet J, Kim JY, Coates NE et al (2007) Nat Mater 6:497
39. Brown PJ, Thomas DS, Kohler A et al (2003) Phys Rev B 67:16
40. Erb T, Zhokhavets U, Gobsch G et al (2005) Adv Funct Mater 15:1193
41. Yang X, Loos J (2007) Macromolecules 40:1353
42. Yang X, Loos J, Veenstra SC et al (2005) Nano Lett 5:579
43. Bjorstrom CM, Bernasik A, Rysz J et al (2005) J Phys Condens Matter 17:L529
44. Jonsson SKM, Carlegrim E, Zhang F et al (2005) Jpn J Appl Phys 44:3695
45. Hoppe H, Sariciftci NS (2006) J Mater Chem 16:45
46. Wang WL, Wu HB, Yang CY et al (2007) Appl Phys Lett 90:183512
47. Hadipour A, de Boer B, Blom PWM (2008) Adv Funct Mater 18:169
48. Hadipour A, de Boer B, Wildeman J et al (2006) Adv Funct Mater 16:1897
49. Dennler G, Prall HJ, Koeppe R et al (2006) Appl Phys Lett 89:073502
50. Gilot J, Wienk MM, Janssen RAJ (2007) Appl Phys Lett 90:143512
51. Kim JY, Lee K, Coates NE et al (2007) Science 317:222
52. Shrotriya V, Wu EHE, Li G et al (2006) Appl Phys Lett 88:064104
53. Brown TM, Friend RH, Millard IS et al (2001) Appl Phys Lett 79:174
54. Brabec CJ, Shaheen SE, Winder C et al (2002) Appl Phys Lett 80:1288
55. Hasegawa T, Miura S, Moriyama T et al (2004) SID Int Symp Dig Tech Pap 35:154
56. Huang JS, Li G, Wu E et al (2006) Adv Mater 18:114
57. Mihailetchi VD, Blom PWM, Hummelen JC et al (2003) J Appl Phys 94:6849
58. Briere TR, Sommer AH (1977) J Appl Phys 48:3547
59. Wang E, Wang L, Lan L et al (2008) Appl Phys Lett 92:033307
60. Svensson M, Zhang FL, Veenstra SC et al (2003) Adv Mater 15:988
61. Inganas O, Svensson M, Zhang F et al (2004) Appl Phys Mater Sci Process 79:31

62. Chen MH, Hou J, Hong Z et al (2009) Adv Mater 21:4238
63. Liao HH, Chen LM, Xu Z et al (2008) Appl Phys Lett 92:173303
64. Winder C, Sariciftci NS (2004) J Mater Chem 14:1077
65. Muhlbacher D, Scharber M, Morana M et al (2006) Adv Mater 18:2884
66. Li G, Chu CW, Shrotriya V et al (2006) Appl Phys Lett 88:253503
67. Scharber MC, Muhlbacher D, Koppe M et al (2006) Adv Mater 18:789
68. Zhang F, Perzon E, Wang X et al (2005) Adv Funct Mater 15:745
69. Zhu Z, Waller D, Gaudiana R et al (2007) Macromolecules 40:1981
70. Dhanabalan A, van Duren JKJ, van Hal PA et al (2001) Adv Funct Mater 11:255
71. Wang XJ, Perzon E, Oswald F et al (2005) Adv Funct Mater 15:1665
72. Campos LM, Tontcheva A, Gunes S et al (2005) Chem Mater 17:4031
73. Lee JK, Ma WL, Brabec CJ et al (2008) J Am Chem Soc 130:3619
74. Zhou QM, Hou Q, Zheng LP et al (2004) Appl Phys Lett 84:1653
75. Lu G, Usta H, Risko C et al (2008) J Am Chem Soc 130:7670
76. Hou J, Chen HY, Zhang S et al (2008) J Am Chem Soc 130:16144
77. Shrotriya V, Li G, Yao Y et al (2006) Adv Funct Mater 16:2016
78. Sylvester-Hvid KO, Ziegler T, Riede MK et al (2007) J Appl Phys 102:054502

Index

A
Absorption coefficients, 192
Adsorbents, 156, 165
AFM. *See* Atomic force microscopy
Air mass (AM), 21
Anisotropy, 139, 140, 144
Annealing temperature, 226
Architecture, 22–30
Atomic force microscopy (AFM), 111, 207

B
Bandgaps, 9, 171
Bathocuproine (BCP), 24
Benzoselenadiazole (BSeD), 46, 64
Benzothiadiazole (BT), 54, 55
Bidentate, 85–90
Bilayer, 27, 70
Bilayer heterojunction, 23
Binding energy cutoff, 181
Binuclear, 97–100
Birefringence, 146
Bright-field TEM, 216
Bulk heterojunction, 26–28, 200
Bundling, 161, 162

C
C_{60}, 183
Capillary force, 160
Carrier injection, 113
Carrier transport, 211
Cationic metal complexes, 108
Cationic transition metal complexes, 107–133
CBP, 148
CCT. *See* Correlated color temperature
CEM. *See* Cutoff emission measurement
Challenges, 30–31
Characteristic penetration depth, 191
Charge collection, 201
Charge collection efficiency, 25
Charge injection, 71
Charge-injection efficiency, 159
Charge mobility, 212
Charge transport, 201
Chemical modification, 165
Chenodeoxycholate, 166
Chromaticity coordinates, 4, 5
Chromophores, 45–50
CIE. *See* Commission Internationale de l'Eclairage
CIE 1931 chromaticity, 82
Colorimetric, 38
Color rendering index (CRI), 4, 6, 39, 88
Color tuning, 118–120
Commission Internationale de l'Eclairage (CIE), 39, 81, 112
Complementary colors, 103
Copolyfluorene, 46
Copolymer, 52
Copper phthalocyanine (CuPc), 183
Correlated color temperature (CCT), 6, 39
Cost, 12
CRI. *See* Color rendering index
Crystallographic structure, 145
CsF, 186
Cutoff emission, 147
Cutoff emission measurement (CEM), 139
Cyclic voltammetry, 118, 119, 124, 128, 229
Cyclometalating, 85–89, 120

D
Degradation, 121, 133
Dexter, 61
Diffusion lengths, 11
Dimer, 68
Dipole, 180, 181
Disorder, 155, 157–159, 164

Donor-acceptor, 18
Dopants, 9
DSSCs. *See* Dye-sensitized solar cells
Dye-dispersed, 57–60
Dye-sensitized solar cells (DSSCs), 154, 158, 165

E
EBL. *See* Exciton blocking layer
Electrochemical cells, 106
Electrochemical gaps, 112
Electroluminescence (EL), 7, 81
Electroluminescent polymers, 65
Electron affinity, 64
Electronic structures, 173, 182
Ellipsoid, 150
Energy, 2
Energy level diagrams, 195
Energy offset, 180
Energy transfer, 9
EQE. *See* External quantum efficiency
Excimer, 68, 85, 87, 92
Exciplex, 68
Excited-state lifetimes, 108, 123
Exciton binding energy, 23, 170–171
Exciton blocking layer (EBL), 23–25, 189
Exciton diffusion efficiency, 19
Exciton diffusion length, 19, 200
Exciton dissociation, 201
Exciton quenching, 24, 122
Excitons, 8
Exponential decay, 123
External quantum efficiency (EQE), 7, 41
Extinction coefficients, 144, 147

F
Fermi-level pinning, 173, 186, 222
Fill factor (FF), 20, 205
Fluorescence, 87
Fluorescent, 63
Fluorescent lamp, 43
Förster, 61
 radius, 116
 transfer, 95
Full-color, 51
Full width at half maximum (FWHM), 55
Future, 13–14

G
GIXRD. *See* Grazing incident X-ray diffraction
Grätzel cell, 154
Grazing incident X-ray diffraction (GIXRD), 207

H
Highest occupied molecular orbital (HOMO), 119, 125
Hole-electron recombination, 91
Hole mobility, 29
HOMO. *See* Highest occupied molecular orbital
Host-guest energy transfer, 130
Hybrid, 11–12

I
IMPS. *See* Intensity-modulated photocurrent spectroscopy
Incident photon to current efficiency (IPCE), 22, 208
Incident solar radiation, 21
Intensity-modulated photocurrent spectroscopy (IMPS), 156
Interchromophore quenching, 116
Interface dipole, 193
Interfacial electronic structures, 178, 183
Interfacial structure, 177
Intermolecular interactions, 110
Internal quantum efficiency (IQE), 41, 83
Inverted polymer solar cells, 220–226
IPCE. *See* Incident photon to current efficiency
IQE. *See* Internal quantum efficiency
Iridium (III), 81
Iridium complex, 66, 122
Isotropic, 142

L
Lamellae, 204
Lamellar, 203
Lateral deflection, 161
LECs. *See* Light-emitting electrochemical cells
Lifetime, 12, 30, 121
Light-emitting electrochemical cells (LECs), 107
Light-emitting layers (EMLs), 6, 7
Lighting, 3, 133
Light source, 13
Low bandgap polymers, 226–234
Lowest unoccupied molecular orbital (LUMO), 119, 125
Luminous efficiency (LE), 40, 88
LUMO. *See* Lowest unoccupied molecular orbital

M
Mesoscopic, 154
Metal doping, 195
Metal-insulator-metal (MIM), 172, 207

Index 239

Metal-to-ligand charge transfer (MLCT), 83
Microcracks, 161
Microstructure, 28
MIM. *See* Metal-insulator-metal
MLCT. *See* Metal-to-ligand charge transfer
Molecular-level bending, 193
Molecular orientation, 138, 140, 210
Monochromatic, 52
Monomer, 88
Morphological disorder, 162
Morphology, 28, 218
Multilayer, 71, 74
Multiple emissive layers, 70–71

N
Nanofibril, 209
Nanoparticle, 157
Nanoporous, 155
Nanostructured, 167
1,8-Naphthalimide, 46
Near-white, 102
Near-white emission, 100
NIR radiation, 28

O
OLED. *See* Organic light-emitting diode
Oligoelectrolytes, 72
Open-circuit photovoltage, 155
Open-circuit voltage, 20, 170, 205
Optical anisotropies, 141
Optical bandgap, 31
Optical interference, 25, 30
Optical out-coupling efficiency, 114
Optical spacer, 25–26, 189, 190
Optoelectronic device, 71
OPV. *See* Organic photovoltaic cells
Organic light-emitting diode (OLED), 80, 84
Organic-organic heterojunctions, 173
Organic-organic interfaces, 182
Organic photovoltaic cells (OPV), 17, 18, 200
Organic photovoltaic devices, 169–195
Organic solar cells, 14–31
Orientational disorder, 160
Oriented, 157–159
Osmium complex, 63
Oxadiazole, 68
Oxidation potential, 119, 128

P
PCE. *See* Power conversion efficiency
π-delocalized, 130
Pendant, 51

Phase segregation, 44
Phase separation, 217, 220
Phase shifts, 142
Phenothiazine, 46, 64
Phosphor, 10
Phosphorescence, 87, 92
Phosphorescent, 10
Phosphorescent-emitting materials, 80
Phosphorescent materials, 44
Phosphorescent polymers, 66
Photoactive, 28–29
Photoluminescence (PL), 205
Photoluminescent, 108
Photons, 17
Photovoltage, 166
Photovoltaic, 17, 223, 228
Photovoltaic response, 19, 190, 191
P3HT:[70]PCBM, 211
Pillow effect, 187
Platinum (II) complex, 48, 84
PL quenching, 232
Polarizability, 146
Poly(p-phenylenevinylene) (PPV), 101
Polyfluorene (PFO), 47, 101
Polymer-based devices, 38
Polymeric, 29
Polymeric materials, 100
Polymer LECs, 107, 111
Polymer light-emitting devices, 44
Polymer solar cell, 234
Polyspirofluorenes, 60
Power conversion efficiency (PCE), 20, 170, 202
Power efficiency (PE), 4, 40, 88, 113
Power-law, 158
Printing technology, 100
Production, 14
Profilometer, 203
PSBTBT:PC$_{70}$BM, 230

Q
Quantum efficiency (QE), 40, 90
Quantum yields, 108
Quinacridone, 46

R
Recombination zone, 114
Reduction potential, 119
Refractive indices, 139
Renewable energy, 3
Root-mean-squared surface roughness, 215
RR-P3HT:PCBM, 202, 234
Rubrene, 57

S

Scanning electron microscopy (SEM), 156
Schiff base complexes, 97
Schottky-Mott model, 172
Self-quenching, 124
SEM. *See* Scanning electron microscopy
Short-circuit current, 20
Short-circuit current density, 205
Short-circuit photocurrent density, 159
Silole-containing polymers, 229
Singlet exciton, 45–48
Slab waveguide, 149, 150
Solar cells, 154
Solar illumination, 227
Sol-gel process, 26
Solid-state lighting, 132
Solid-state organic devices, 31
Solution-processed, 100
Solvent annealing, 202, 215, 219
Spin-coating, 213, 217
Star-shaped, 48
Surface defects, 164
Surfactant, 219

T

Tandem, 29
Tapping mode AFM, 208
TEM. *See* Transmission electron microscopy
Tetradentate, 95–97
TFT. *See* Thin-film transistors
Theoretical limit, 176
Thermal annealing, 51, 224, 230, 231
Thermal crosslinking, 74
Thin-film transistors (TFT), 233
Thiophene, 64
TiO_2, 155
Transition dipole moments, 138
Transition metal complexes, 106
Transmission electron microscopy (TEM), 156, 213
Transport time constants, 158, 163
Tridentate, 90–95
Triplet emitter, 61
Triplet energy, 10, 66, 109
Triplet excited states, 103
Triplet excitons, 48

Turn-on time, 125, 132
Turn-on voltage, 55, 63
Twistacene, 57

U

Ultraviolet photoemission spectroscopy (UPS), 174
  measurements, 222
  studies, 176–178

V

Variable angle spectroscopic ellipsometry (VASE), 138
VASE. *See* Variable angle spectroscopic ellipsometry

W

White emission, 38
White inorganic light-emitting diodes (WLEDs), 2
White light, 61, 127
White light source, 42
White organic light-emitting devices (WOLEDs), 2, 3
White organic light-emitting diode (WOLED), 80, 83, 87, 94, 106
White polymer light-emitting devices (WPLEDs), 57–60
WLEDs. *See* White inorganic light-emitting diodes
WOLED. *See* White organic light-emitting diode
WOLEDs. *See* White organic light-emitting devices
Work function, 187, 223
WPLEDs. *See* White polymer light-emitting devices

X

XPS. *See* X-ray photoelectron spectroscopy; X-ray photoemission spectroscopy
X-ray diffraction (XRD), 144, 156
X-ray photoelectron spectroscopy (XPS), 213
X-ray photoemission spectroscopy (XPS), 174
XRD. *See* X-ray diffraction